T0419336

EARTH SCIENCES IN THE 21ST CENTURY

SAND DUNES: CONSERVATION, TYPES AND DESERTIFICATION

EARTH SCIENCES IN THE 21ST CENTURY

Additional books in this series can be found on Nova's website under the Series tab.

Additional E-books in this series can be found on Nova's website under the E-books tab.

EARTH SCIENCES IN THE 21ST CENTURY

SAND DUNES: CONSERVATION, TYPES AND DESERTIFICATION

JESSICA A. MURPHY
EDITOR

Nova Science Publishers, Inc.
New York

For permission to use material from this book please contact us:
Telephone 631-231-7269; Fax 631-231-8175
Web Site: http://www.novapublishers.com

Library of Congress Cataloging-in-Publication Data

Sand dunes : conservation, types, and desertification / editor, Jessica A. Murphy.
p. cm.
Includes bibliographical references and index.
ISBN 978-1-61324-108-0 (hardcover)
1. Sand dune conservation. 2. Sand dune ecology. 3. Desertification. I. Murphy, Jessica A.
QH88.5.S26 2011
577.5'83--dc22

2011008480

Published by Nova Science Publishers, Inc. †New York

CONTENTS

PREFACE

In this book, the authors present current research in the study of the conservation, types and desertification of sand dunes. Topics discussed include the aridization, dune dissipation and pedogenesis in the Quarternary of Eastern Pampean sand sea; desert sand dunes for sulfur concrete production; Allocosa brasiliensis as a model towards the conservation of coastal sand dunes in Uruguay and the causes, impacts and control of desertification.

Chapter 1 - The Pampa region is a large plain of aeolian origin which forms the central part of Argentina. After a warm and humid environment established during the Last Interglacial, a cold and dry climate provoked by the glaciations in the Andes Cordillera generated a desert formed by a sand sea (the Pampean Sand Sea, PSS) and loess deposits (the Peripherial Loess Belt, PLB). Later climatic oscillations produced reworking of the sand structures, with destruction, reshaping and dissipation of dunes, pedogenesis and new deflation of the soil. Such a sequence occurred several times in the Pampa. This chapter describes the processes and dune types of the northeast of the PSS, with an area covering some 10,000 km^2. A sequence of six desert climates intercalated with humid or subhumid pulses were recognized in the region.

Three basic processes were responsible for the evolution of the sand sea: deflation, dune dissipation and pedogenesis, each of them recognizable in morphology and internal sedimentary structures. Deflation was provoked by strong winds originated in the Patagonian ice cap, that formed longitudinal mega-dunes during the Oxigen Isotopic Stage 4 (OIS 4) and other dune types in OIS 2 and Late Holocene. Dune dissipation, a process produced by the pluvial action on loose sand in semiarid climates, was particularly important in the climatic change occurred during OIS 3 (between 65 and 36 ka. BP). Pedogenesis dominated during the middle Holocene Hypsithermal and a Late Holocene (3,5-1,4 ka. BP) dry climate deflated again the PSS, generating composed parabolic mega-dunes.

Typical weather scenarios of the Holocene past climates are sporadically reproduced today during extreme humid or dry short periods (one year or two), which allows to reasonably know the basic meteorological parameters of those past systems: Dry periods similar to that of Late Holocene are characterized by a large thermal amplitude, frosts and stronger winds than normal, reproducing a continental anticyclonic circulation. Humid extremes similar to the Holocene Optimum Climaticum are warmer than normal, with lower thermal amplitude, with rains produced by local convection processes.

Chapter 2 - The strength and stability of desert sand dunes present challenging problems to geotechnical engineers. To overcome these problems, sulfur concrete as a new material was developed and evaluated in view of its microstructure, physical, thermal, mechanical, hydraulic, and chemical properties. Sulfur is used as the main binding material in sulfur concrete for years. The use of elemental sulfur with aggregates to manufacture sulfur concrete encountered problems due to the complex nature of sulfur chemistry. While initially excellent strength properties were obtained, over a short period of time the material would fail. However, eventually it will convert to a brittle orthorhombic crystal structure with poor strength characteristics. This study has focused on the modification of elemental sulfur; the key to this modification came with the discovery of a novel technique to treat sulfur by preventing the allotropic transformation. The sulfur concrete was prepared from modified sulfur, elemental sulfur, fly ash, and desert sand dunes (soil aggregates). Sulfur concrete specimens were cured in air, water, acidic and saline solutions at different temperatures, ranging from room temperature to 60° C, at different periods. The results indicated that the strength of the manufactured sulfur concrete is about three times higher than that of Portland cement concrete. The results of hydraulic conductivity are ranged between 1.46×10^{-13} and 7.66×10^{-11} m/s making it a good candidate for its potential use as barrier system in arid lands. Durability results indicated that the manufactured material has high resistance to chemical environments being acidic, neutral, and saline and extremely low leached sulfur from the solidified matrix was recorded.

Chapter 3 - *Allocosa brasiliensis* is a wolf spider that constructs burrows along the coastal sand dunes of Uruguay. They are medium-sized nocturnal spiders with whitish coloration that turns them cryptic with their sandy habitat.

The species shows a reversal in both sex roles and sexual size dimorphism expected for spiders, being the first case reported in Araneae. Females are the mobile sex that looks for sexual partners and initiates courtship, and males are larger than females. Females prefer to mate with males showing deeper burrows. Copulations occur inside the male burrows and, after copulation ends, males donate their own burrows to the females, which officiate as nests for the future progeny. The ability of digging deep burrows in the sand is an adaptation to survive in the harsh coastal ecosystem where these spiders face drastic temperature variations, excessive heats, strong winds and prevent water loss. The atypical sex strategies found in this wolf spider could be related to ecological constraints. Prey availability fluctuates, showing high variability in quantity and quality, and high levels of cannibalism were observed. During the last decades, the Uruguayan coastline has been dramatically reduced and modified due to the introduction of exotic flora, urbanization and tourism, what has lead to habitat reduction and isolation of *A. brasiliensis* populations. The strict association between this spider and the sand dunes, in addition to its vulnerability to human impact and role as a generalist carnivore, turn this wolf spider into a good candidate for terrestrial biological indicator of conservation in coastal ecosystems. Behavioral, evolutionary and ecological studies on this species are essential for implementing adequate management and conservation plans for these areas.

Chapter 4 - The eastern coastline of Australia became subjected to intensive mining activity in the 19th and 20th centuries, which has resulted in disturbed sand dune systems. Public pressure has subsequently led to some protection measures and restoration programs in disturbed sand dune areas implemented with varying degrees of success The re-modelled surficial landscape resembles the surrounding landscape in that local species have been planted to increase the success rate. Mine site rehabilitation on Fraser Island, now a World Heritage Site, however, has produced poor biodiversity-outcomes over time. This is reflected in species-poor post-disturbance plant communities that lack key-stone species and any structural heterogeneity to provide critical wildlife habitat. Understanding the occurrence, distribution and the role of the soil-biotic factors involved in the natural recovery process may therefore benefit the revegetation and restoration of disturbed sand dune systems. The chapter presented here questions whether disturbance of healthy actinomycete and mycorrhizal populations in the course of sand mining, topsoil stripping, storage and re-application is among the reasons for poor recovery response observed on the island over the last 30-40 years.

Chapter 5 - The Quaternary sedimentary basin of Senegal is marked by phases of edification of varied dune systems, differently distributed in time and space. Three systems are identified. Akcharien erg, prior to 100 ka BP, leveled or more or less dismantled in places. Ogolian erg built along the Atlantic coast of Senegal and Mauritania during the arid phases of Würm III. Barrier beaches over Holocene and Present, built during the regularization of shorelines. These dune systems mainly contain quartz and useful heavy minerals relatively wellpreserved in the barrier beaches. Climate change of the region and eustatic fluctuations have led to a particular pattern of Ogolian sand dunes conducive to the installation of a groundwater aquifer and lush vegetation in areas now transformed into interdunes market-gardening. The drought of the 70s and anthropogenic activities have led to degradation of ecosystems including the marked decline of groundwater levels, deforestation, remobilization of dune fronts and sanding up farming areas.-

Chapter 6 - Desertification is a serious environmental problem and it potentially affects 35% of the land surface of earth and 32% of the human population. Desertification is land degradation in arid, semi-arid and dry sub-humid areas and includes degradation of vegetation cover, soil degradation, and nutrient depletion. Overcultivation, increased fire frequency, overdrafting of groundwater, livestock grazing, deforestation, water impoundment, poor irrigation management, increased soil salinity, and global climate change are the main causes of desertification. The different processes involved in desertification include wind erosion, soil erosion, salinity-alkalinity, and waterlogging. Africa, Asia, Latin America, and the Caribbean are the most regions threatened by desertification. The impacts of desertification include environmental impacts, economic impacts, and poverty and mass migration. A number of methods have been used in order to reduce the rate of desertification. These methods include restoring and fertilizing the land, reforestation, developing sustainable agricultural practices, and the traditional lifestyles.

In: Sand Dunes
Editor: Jessica A. Murphy, pp. 1-42
ISBN 978-1-61324-108-0

Chapter 1

ARIDIZATION, DUNE DISSIPATION AND PEDOGENESIS IN THE QUATERNARY OF EASTERN PAMPEAN SAND SEA

Martin Iriondo[1*], Daniela Kröhling[2] and Ernesto Brunetto[3]
[1]CONICET ; CC 487 (3100) Paraná, Argentina
[2]CONICET – Universidad Nacional del Litoral; CC 217 (3000) Santa Fe, Argentina
[3]CICyTTP-CONICET; (3105) Diamante, Entre Ríos, Argentina

ABSTRACT

The Pampa region is a large plain of aeolian origin which forms the central part of Argentina. After a warm and humid environment established during the Last Interglacial, a cold and dry climate provoked by the glaciations in the Andes Cordillera generated a desert formed by a sand sea (the Pampean Sand Sea, PSS) and loess deposits (the Peripherial Loess Belt, PLB). Later climatic oscillations produced reworking of the sand structures, with destruction, reshaping and dissipation of dunes, pedogenesis and new deflation of the soil. Such a sequence occurred several times in the Pampa. This chapter describes the processes and dune types of the northeast of the PSS, with an area covering some 10,000 km^2.

* E-mail: martiniriondo42@yahoo.com.ar

A sequence of six desert climates intercalated with humid or subhumid pulses were recognized in the region.

Three basic processes were responsible for the evolution of the sand sea: deflation, dune dissipation and pedogenesis, each of them recognizable in morphology and internal sedimentary structures. Deflation was provoked by strong winds originated in the Patagonian ice cap, that formed longitudinal mega-dunes during the Oxigen Isotopic Stage 4 (OIS 4) and other dune types in OIS 2 and Late Holocene. Dune dissipation, a process produced by the pluvial action on loose sand in semiarid climates, was particularly important in the climatic change occurred during OIS 3 (between 65 and 36 ka. BP). Pedogenesis dominated during the middle Holocene Hypsithermal and a Late Holocene (3,5-1,4 ka. BP) dry climate deflated again the PSS, generating composed parabolic mega-dunes.

Typical weather scenarios of the Holocene past climates are sporadically reproduced today during extreme humid or dry short periods (one year or two), which allows to reasonably know the basic meteorological parameters of those past systems: Dry periods similar to that of Late Holocene are characterized by a large thermal amplitude, frosts and stronger winds than normal, reproducing a continental anticyclonic circulation. Humid extremes similar to the Holocene Optimum Climaticum are warmer than normal, with lower thermal amplitude, with rains produced by local convection processes.

1. Introduction

Large dune fields, or sand seas (ergs), are landscapes often thought to be found only in deserts beneath the great, subtropical high-pressure zones, where subsiding air suppresses rainfall. Dune fields are also quite common in mid-latitude regions, to the north and south of subtropical deserts (Sun and Muhs, 2007). Wilson (1973) analyzed the distribution of the areal extent of aeolian sand bodies and found a sharp break in the size distribution of collections of dunes at about 32,000 km^2 , discriminating a larger group with a sharp peak in size at about 200,000 km^2. Most dunefields and sand seas occur in basins or lowlands, and many sand seas nearly fill their basins (Warren, 2006).

According to the analysis of the largest areas of stabilized dunes in the midlatitudes, Sun and Muhs (2007) cited for South America the Pampas region of Argentina. On the contrary, Lancaster (2007) indicated that large sand seas are absent in the Americas and refers the existence of low-latitude dune fields, between them the vegetation-stabilized relict dune fields of the northern Pampa. The Pampean Sand Sea (PSS) is the largest sand sea in South America and one of the largest in the world and covers continuously a large area. It is a non-tropical desert body, that is, the origin and development of the system is not directly produced by the general circulation of the atmosphere, but by major complex geological factors. The PSS is a mid-latitude large dune fields in the sense of Sun and Muhs (2007); the difference respect to the lower latitude sand seas is that many of those in mid-latitudes are in semiarid, rather than arid climates, and therefore are not presently active (interpreted as aeolian sand bodies that are not covered with vegetation, and where particles are currently being transported by the wind). Where sand is vegetated and particles are not being transported by the wind, a sand body is 'stable.' Table I.

Table I. Comparison of The Pampean Sand Sea with other eolian systems; a) in South America (own data); b) in the World (from Mc Kee 1979 and Short and Blair 1983)

Sands seas in South America	Surface (km^2)
Pampean Sand Sea	**200,000**
Orinoco (Venezuela)	60,000
Río Branco (Brazil)	40,000
Piura (Peru)	15,000
Sao Francisco (Brazil)	8,000

World sands seas	Surface (km^2)
Rub al Khali (Arabia)	560,000
Great Sandy Desert (Australia)	360,000
Simpson Desert (Australia)	300,000
Takla Makan (China)	261,000
Pampean Sand Sea	**200,000**
Grand Erg Oriental (Algeria)	192,000
Kalahari (South Africa)	100,000
Thar (India/Pakistan)	100,000
Grand Erg Occidental (Algeria)	80,000
An Nafud (Arabia)	72,000
Marzuq (Morocco)	58,000
Tengger (Mauritania)	42,700
Namibia (Namibia)	34,000
Ala Shan (China)	33,000
Mu Us (China)	32,000
Wahiba (Arabia)	16,000
Lut (Iran)	10,000

Probably, the most interesting characteristic of the PSS consists in being a non-permanent desert, and undergoing relatively long humid periods; in fact, one of such events is the present climate of the region. That allows the possibility of register several phases of formation (six) and destruction (also six) of desert landscapes. Owing to the large extension of the region and the particular grain size composition of the sediments (basically very fine silty sand), deposits are very sensitive to minor environmental changes. Such conditions marked significant differences among successive "reconstructions" of the dune fields. Obviously, some methods of Field Geology must be adapted and other routines, particularly those developed for Hydrogeology and Geotechnics, employed in the Quaternary stratigraphy of the sand sea. This work is centered on the estern half of the PSS. Important advances on the research of sand seas were produced by Wilson (1973), McKee (1979), Goudie (2002) and Lancaster (1999). In modern sand seas, there is an increasing body of evidences that suggests that they have accumulated episodically (Talbot, 1985, Kocurek et al., 1991, Lancaster, 1993).The construction of the sand seas has been determined by climatic, tectonic, and sea-level changes that have affected sand supply, availability, and mobility, as well as the preservation of deposits and landforms from prior episodes of aeolian construction. Many sand seas contain a variety of dune types or dunes on different trends, often forming complex patterns (Lancaster, 2007). Warren (2006) cited the following reasons that explain the variety of dunes in a sand sea: 1) variety is a function of size, for different wind regimes can occur over such large areas, and wind regime is a major determinant of dune type, 2) variety of dune form is a function of age. Large bodies of sand can only accumulate over many thousands of years, span of time that have seen changes in climate. Older dunes may differ in type and orientation from younger ones. Moreover, many older dunes have been subdued by erosion, and have developed soils at times when the climate was wetter. Their gentle topography may be scarred by haphazard reactivation or it may be buried by younger sands to varying degree and 3) variety of dune type in large sand seas is a function of the movement of the dune body as a whole.

Pampean Sand Sea is particularly sensitive to climatic changes, mainly to shifts in the precipitation/evapotranspiration balance. Shifts to drier conditions produced the re-activation of dunes fields (dune building events or remobilization of previus aeolian sand deposits) and the generation of aeolian erosive landforms, while shifts to humid conditions stabilized dunes (developing of stable surfaces, which would be marked by paleosols) and generated ponds or shallow lakes in interdune areas or in the deflation hollows or corridors. Deflation of existing dune areas and sand sheets occurred in many parts of the sand sea, mainly during subhumid phases. Shifts of the climatic belts on the PSS during the Late Quaternary left a complex pattern of stabilized and reactivated field dunes of different types. Local processes as dune dissipation, salt weathering and dynamics of superficial water masses in humid periods revealed to be of major importance. And not least, after twenty years of sieving dunes and megadunes of different types, the sand-silt limit in grain size, when one tries to define both categories on physical (not mathematical) properties and behaviour, remains still in discussion.

The age of the aeolian sedimentary units dated by OSL and TL and also the units which age may be inferred from stratigraphic correlation with other Quaternary units, is referred here to the oxygen-isotope stages (OIS).

Applying the principle that extreme meteorological conditions in the present climate reproduce normal synoptic climatic structures of past climates, several climatic parameters not preserved in sediments are reconstructed in this chapter. Related disciplines, as Palinology and Paleobotany, use customarily this kind of theoretical basis, reconstructing past climates based on the ecology of living plants.

2. Location and Dimensions

The Pampean Sand Sea forms an irregular body which covers most of the South Pampa in an extension of 130,000 km^2. The extreme geographic points of this area are 33° and 38° lat.S and 59° 20' and 67° long. W. The thickness of the sand is between 12 and 20 meters in the eastern half and probably more than 30 m in the west. It is the largest sand sea in South America (Iriondo, 1992; Iriondo and Brunetto, 2008; Petit-Maire et al., 1999) and one of the largest in the World, according to the figures published by Mc Kee (1979) and Short and Blair (1986). The comparison of different systems appears in Table I. (Figure 1).

3. Methodology

The methodology applied in the study of the Quaternary of the Pampa integrates several different disciplines of Geomorphology and Sedimentology, in order to have a multi-proxi set of data.

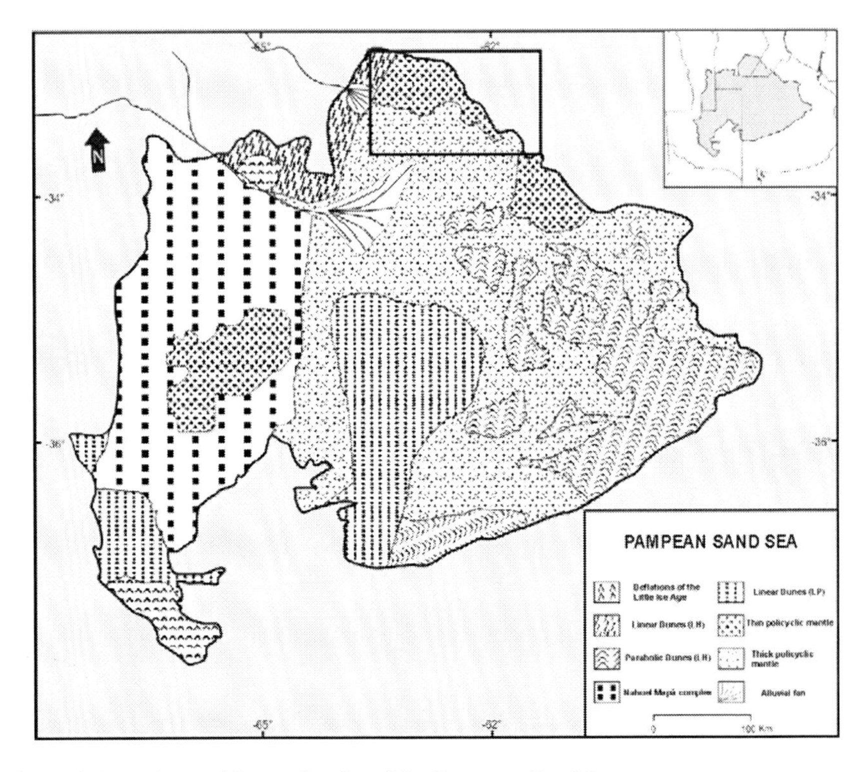

Figure 1. Location and internal units of the Pampean Sand Sea.

The work included field, laboratory and computer techniques, both classical and advanced. The geological and geomorphological cartography was produced by our working group (Iriondo, 1987, 1992; Petit-Maire et al., 1999). Local details were complementary mapped by photo-interpretation and topographic quadrangles in 1:50,000 scale. Satellite image analysis, combined with research on aeolian geomorphology and sedimentology were used to analyze the dune patterns and identify genetically distinct generations of dunes in order to advance in the reconstruction of the Quaternary history of the northeastern of the sand sea.

Morphometric analysis of Geographic Information Systems (GIS) was applied by using GRASS-GIS (GRASS Development Team, 2005), following the methodology proposed by Grohmann et al. (2005), in order to obtain hypsometry, slope, aspect (slope orientation) and surface roughness maps. Also complementary swat profiles were obtained. The basic element for deriving morphometric maps is the digital elevation model (DEM), available from Shuttle Radar Topographic Mission (SRTM, distributed at horizontal resolution of 3arcsec approximately a 90×90 m grid-, Jarvis et al., 2008). The DEM was smoothed with a 3x3 filter to minimize the effects of possible noises. This kind of maps and profiles provided a broader view of altimetric behavior, and helped to determine inclination of large topographic features in plains. The hypsometric and the aspect (slope orientation) maps were the best in the case of the Pampean Sand Sea, the first one for marking clearly the limits of the Late Holocene megadunes and the aspect map because it reveals a trend produced by the last movement of sand occurred in the Little Ice Age.

Field work was performed during twenty years. The first expedition to the region was made in 1985, during a dry interannual period; that facilitated a good description of outcrops in river banks, old quarries, lunettes and dry dell bottoms, besides deflation and sand movement. During the last decade (characterized by humid years) were developed three major expeditions, during which the dynamics of water in shallow lakes and on the landscape could be technically described. The work included systematic sampling of dune fields and drilling of seven 20 meter deep boreholes by rotation drill in key locations of the system, and several auxiliary holes using manual auger down to 4 meters. Undisturbed samples at each meter of advance were taken in the larger boreholes by using geotechnical standard methods (SPT: Standard Penetration Test) with recovering of internal samples.

The SPT is based on the penetration of a tube with standard diameter (35 mm) by hitting with a standard hammer (75 Kg weight) and counting the number of hits needed for advancing 75 cm (Terzaghi and Peck, 1963). This method, although classical in Soil Mechanics, is not exploited in regional Quaternary studies, in spite of its clear advantage to personal opinions. The point is that compaction is an important parameter, frequently used in description and discrimination among Quaternary formations with similar lithologies; such descriptions are normally made with free criteria and in a broad qualitative mood. Hence, that depends on the observational skill of the field geologist; although very useful for each case, the comparison is not valid in general. The SPT method provides really important information. As a complementary measurement, the velocity of advance of the tool provided interesting results in the identification of contacts and thickness of different sandy formations in the subsoil.

The location of the boreholes was preceded by geoelectric exploration by means of EVS (Electrical Vertical Sounding). Owing to the small diameter and the scarce variability of the sand grains in the PSS, the ro-tap analysis was made with intervals of ¼ phi between 125 and 37 μm, following the Udden-Wentwort scale.

Particle size analysis of cores were carried out by contrasting methods: a) by sieving for fractions > 63 μm and sedimentation for fractions < 63 μm, b) by sieving for fractions > 37 μm, for intervals of ¼ φ, c) by Low Angle Laser Light Scattering (LALLS) using a Malvern Mastersizer S. The mineralogical analysis was made in microscopic loose-grain routine, with previous extraction of iron and carbonate films. Besides, X-Ray diffraction analysis was performed on total samples (Cu-Kα radiation). Complementary was made the morphological description of grains; the shape (roundness and sphericity) was measured in the 74 μm fraction, by visual comparison in the "Chart of Visual Estimation of Roundness and Sphericity" (Krumbein and Sloss, 1955). Surficial textures of grains were also studied in the same grain fraction, following in general the Krinsley and Margolis technique (in Carver, 1971). The application of luminescence techniques for datings periods of dune formation and/or reactivation contributed to the environmental reconstruction of the region.

Climatic Scenarios

Present Climate

The present climatic pattern in the Southern Hemisphere is controlled by the general circulation of the atmosphere in a rather simple pattern, owing to the dominant oceanic surface of the whole system. The ocean is interrupted by relatively small continental masses located at regular intervals. One of them, South America, can be visualized as a low-lying terrain bordered by a high mountain chain at the western fringe, the Andes Cordillera. This is the unique major feature that influences the atmospheric general circulation.

Three major atmospheric systems occur on South America: The Intertropical Convergency Zone (ITCZ), the Trade-winds and the Westerlies. The ITCZ migrates to the north in July and to the south in January, during the southern summer. During winter, the southern Trade-wind enters deeply in the Amazonia and turns to the south at the foot of the Andes, losing humidity (Clapperton, 1992). Farther on, between the latitudes of 18 and 28°, it is transformed in a dry wind that generates dune fields in the Western Chaco of Bolivia and Argentina. The Westerlies cross the Andes Cordillera in Patagonia, precipitating snow and rain in the mountains, and a desert of the type named "topographic shadow" appears to the east; under particular conditions, the air mass circulates to the northeast-north with anticyclonic curvature, reaching the Pampa in central Argentina. That cold and dry wind is named “Pampero”, it dominates the Pampean weather in dry years and generated the Pampean Sand Sea during the OIS 4.

In addition, the South American Low-Level Jet (SALLJ) is a relevant feature of the warm season low-level circulation and represents a poleward transport of warm and moist air concentrated in a relatively narrow region, with strong wind speeds at low levels downstream and to the east of mountain barriers. In southern South America, episodic incursions of mid-latitude air to the east of the subtropical Andes (also referred to as South American cold surges) are a distinctive year-round feature of the synoptic climatology (Garreaud, 2000). Also, South American cold surges dominate the synoptic variability of the low-level circulation, air temperature, and rainfall over much of the continent to the east of the Andes. The large-scale environment in which the cold air incursion occurs is characterized by a developing mid-latitude wave in the middle and upper troposphere, with a ridge immediately to the west of the Andes and a downstream trough over eastern South America. At the surface, a migratory cold anticyclone over the southern plains of the continent and a deepening cyclone centered over the southwestern Atlantic grow mainly due to upper-level vorticity advection.

At both sides of South America, semi-permanent oceanic anticyclones dominate the tropical latitudes. A second-order anticyclone occurs on the continent, over the Chaco and Pampa regions. It is smaller and weaker than the oceanic ones, and very sensitive to climatic changes and interannual variations of climate. At present times, it takes form only sporadically during a few days and fades out; it is more stable in the upper atmosphere. Only during abnormally dry years, it remains months-long on the surface.

The precipitation gradient in the Pampa region is steep, with present mean annual precipitation ranging from less than 500mm (southwest) to more than 1,000mm (northeast).

4. THE PAMPEAN AEOLIAN SYSTEM

The Pampean Aeolian System covers more than 600,000 Km^2 in central Argentina. It is composed of a sand sea (PSS) and a peripherial loess belt (PLB) (Iriondo, 1997). The sand sea, with more than half of the total surface, extends from the latitude of 33 to 37° S. The sediment deposit is thin in the eastern portion, between 5 and 15 meters, and up to 30 m at the western fringe. Most of the sand is fine to very fine, yellow to yellowish brown in color. The mineral composition is variable, owing to three different sources of material. The main source were the products of physical weathering of the central Cordillera (27/36° lat.S), a high mountain complex characterized by fine-grained Tertiary rocks (Tullio, 1981; González, 1983). A second, regionally restricted source area, were the Pampean Ranges that formed a several kilometers wide fringe of plutonic- and metamorphic-derived sand (Latrubesse and Ramonell, 1990; Cantú, 1992). The third source was the volcanic ash produced by Patagonian volcanoes, which is particularly abundant in the Buenos Aires province, in the southwest of the system (González Bonorino, 1965; Zárate and Blassi, 1990).

The southwestern area of the PSS (northwestern Buenos Aires province and southern San Luis province) is dominated by large longitudinal dissipated dunes (over 200 km long and 2–5 km wide), arranged as extensive southwest-to-northeast archlike. Theses longitudinal dune forms are easily recognizable on satellite images, although the low relief of these landforms makes them difficult to identify in the field. The dunes are presently stabilized by vegetation (grasses and crops). Longitudinal dunes are Late Pleistocene (early last glacial, or OIS 4) age (Iriondo, 1999). The largest longitudinal dune field is located at the center of the system; the dunes of that area are particularly well preserved and have a gentle anti-clock curvature, typical of anti-cyclonic circulation in the Southern Hemisphere. The interdune area is in general occupied by ponds or shallow lakes, characterizing a landscape with a poorly organized drainage (Figure 2).

The northeastern part of the sand sea (San Luis, Córdoba and Santa Fe provinces) exhibit a complex pattern of dune construction, reactivation, and stabilization, with different types of dunes and erosive aeolian landforms. Fields of parabolic dunes, most of them also stabilized and oriented in a southwest-to-northeast direction are common, in general developed from reworking of previous dunes. The limit of the sand sea follows an irregular SE-NW direction through some 800 kilometers in the provinces of Buenos Aires, Santa Fe, Córdoba and San Luis. Behind a 5 to 15 Km wide transition belt, the loess covers the remaining Pampa and neighboring areas. It is composed of friable silt and loam, yellowish brown in color, with frequent concretions of calcium carbonate. Illite is the dominant clay mineral. The original thickness of the formation is 5 to 10 meters, with a record of 52 m in north Santa Fe (Kröhling, 2008).

According to the available information, the Pampean Aeolian System was generated at the beginning of the Late Pleistocene as a result of the glaciations in the Andes.

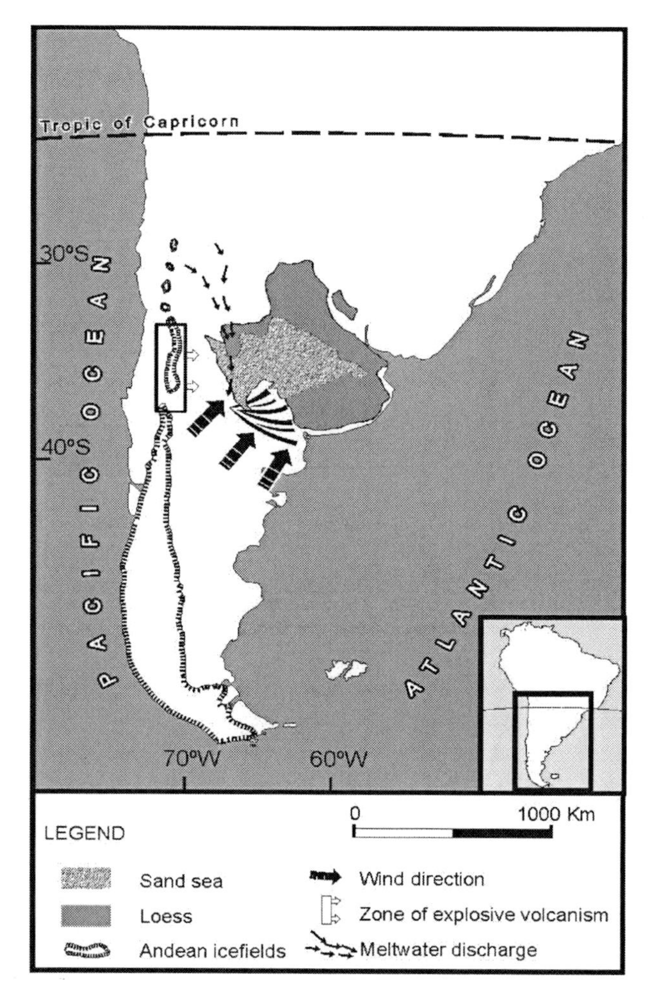

Figure 2. Origin of the Pampean Sand Sea during the IS4.

The ice field covered a large area south of 28° latS (Clapperton, 1994), allowing the strong influence of the South Pacific Anticyclone and producing cathabatic winds blowing toward NNE, registered in the dune fields. The rest of the Argentine Cordillera, north of 28° lat S, was almost free of ice, owing the dryness of the climate. Such a periglacial environment very efficiently produces silt, very fine sand, and illite through physical weathering.

The sediment was supplied to source areas during periods of increased fluvial activity, mainly transported to the south along the Cordilleran piedmont by the Bermejo-Desaguadero-Salado fluvial system. It is a large fluvial net, which covers 248,000 Km^2 in western Argentina. At present, it is disintegrated and basically inactive, owing to the desertic climate of the area; in more humid periods of the Quaternary the collector channel conveyed large discharges. The basin is formed by important Cordilleran tributaries (the rivers Jáchal, San Juan, Mendoza, Tunuyán and Atuel) with significative glacial and nival processes in their headwaters, composed of volcanic rocks (Ahumada, 1990). The products of nival weathering accumulated in maraines and related deposits are rich in silt and fine sand fractions. The rivers formed periglacial terraces in the valleys and wide alluvial fans in the piedmont zone; the Tunuyán and Atuel fans cover areas larger than 10,000 km^2. The whole scenario suggests a large production of fine sediments.

The collector of the fluvial net flows southward for 1,000 km from 28°30' lat. S to 37° 30' lat. S. It is at present partially covered by the major alluvial fans and by Holocene sand fields. In the province of La Pampa it formed a 25-35 km wide flood plain during the upper Pleistocene, filled with more than 35 meters of interestratified silt and sand layers. The mineral association of the sediment indicates a composition similar to those of the Cordilleran headwaters (Salazar, 1983). That large river formed during a humid climate in the Pleistocene was active during the Last Glacial Maximum, very probably because the snow precipitation in the high Cordillera was larger than today. González (1981) found that the maximum lake level in Salina del Bebedero (a tectonic depression connected with the Desaguadero) occurred about 18,000 years before the Present. Based on mineralogical data, González concluded that the sediments of the lake originated in the Atuel basin.

Once the sediments transported by floods of meltwater arrived to a latitude of 37/38° S became available for transport in subsequent drier episodes being deflated by SSW wind (Pampero) coming from the Patagonian ice field. In that region, near to the northern border of the Patagonia, the sand sea began to develop in cold desert climate. The fields of longitudinal dunes make evident the desert climate and the supply of sand. To the NE the climate was peridesertic, allowing the sedimentation and fixing of the dust transported in suspension by the wind, and forming the loess mantle.

5. Local Processes

In order to understand the genesis and evolution of the Pampean Sand Sea, several interrelated local processes should be analyzed. Such processes and mechanisms are often disregarded in the literature, in spite of their real importance in the dynamics of desert and peri-desert landscapes. The most important of them are the following:

5.1. Types of Megadunes

Four types of megadunes were generated in the eastern half of the Pampean Sand Sea during the Late Pleistocene and Holocene times: longitudinal (linear), fish-scale, parabolic and arcuate.

**Longitudinal megadunes* were formed during the first phase of evolution of the system (OIS 4), those structures appear in a very regular pattern, with SSW-NNE and S-N directions, drawing a slight anti-clockwise curve. Such bodies have individual lengths from 50 to 200 km, distances between 3 and 5 Km between successive crests and 5 to 7 m relative heights. In detail, simple linear dunes occur as straight, length, parallel and regular spacing bodies, with a high ratio of dune to interdune areas. The field of longitudinal megadunes is visible only in satellite images and aereal photographs; dissipation processes transformed the original relief (probably several dezens of meters) in a flat surface. The original extension of the field covered about 100,000 km^2; a typical area of occurrence is located around General Villegas (35° lat. S-62°30' long.W) (Figure 3).

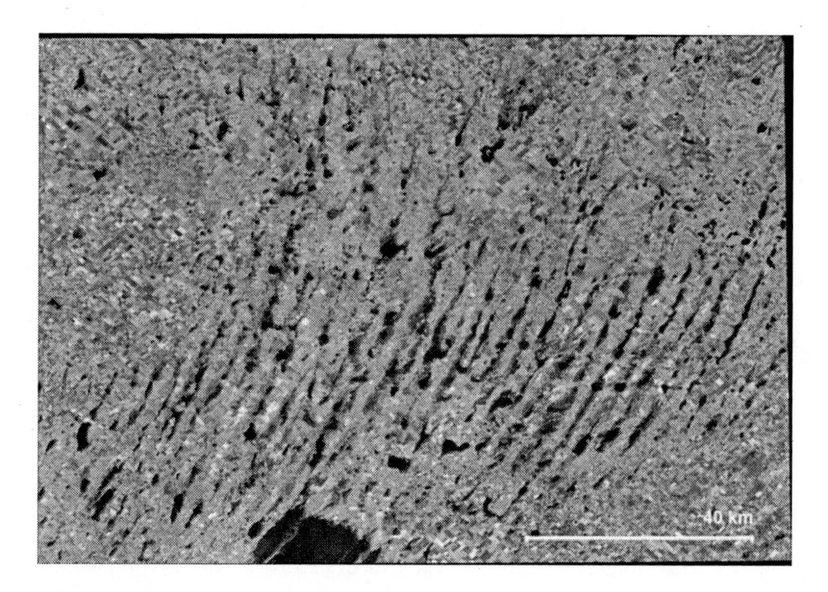

Figure 3. Longitudinal megadunes at Pehuajó.

This type of megadunes outcrops in excavations and quarries covered in discordance by the Little Ice Age loose sands. Megadunes are formed by sets with different resistence to the erosion owing to differences in grain size and carbonate epigenesis. Sets have sub-horizontal position and are composed of medium strata (15 to 40 cm thick), which are internally laminated. Such a condition indicates probable weak winds with scarce turbulence.

The landscape in these areas is characterized by very smooth long slopes and completely flat sectors.

**Fish-scale megadunes* were formed in a 7600 km^2 field in SW Córdoba and have smooth semicircular to triangular shape, pointing to the north. They are remarkably regular in form and size in all the area, 5 to 7.5 Km long and 4 to 6 Km wide and are randomly located inside the field. The structures are Late Holocene in age and are still well preserved and visible in the field. Thicknesses at present times (after dissipation) reach values of five meter or less. The typical locality for this type of dune is Canals (33°40' lat. S-62°40' long. W) (Figure 4).

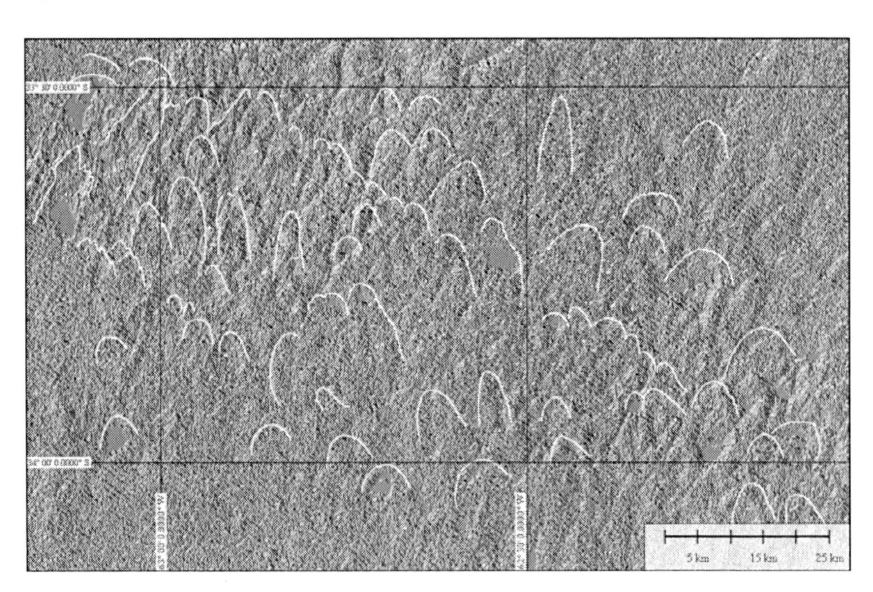

Figure 4. Fish-scales megadunes south of Canals.

**Parabolic megadunes* are irregular complexes of aeolian sand, which advance at different rates in internal sets with triangular shape. Such complexes were mapped south of La Carlota (33°20' lat. S-63°20' long. W) (Figure 5).

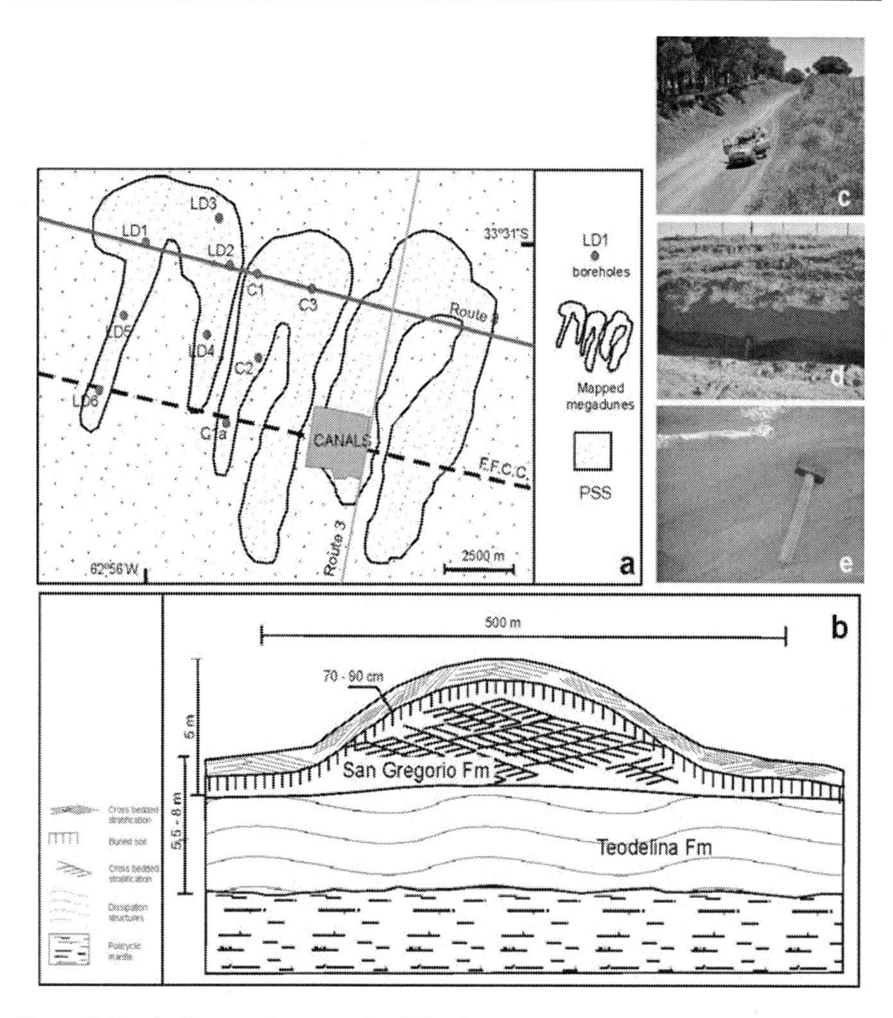

Figure 5. Parabolic megadunes north of Canals.

These dunes are characterized by a U shape with two partly vegetated arms that trail up wind. It will occur in areas of partial vegetation cover and similar wind regimes and tend to migrate downwind, at a rate that is inversely proportional to their height (Goudie, 2006).

**Arcuate megadunes* form long series of archs, each of them 10 to 15 Km long and only 300 to 500 meters wide, separated by depressions with the same dimensions. Typical series have 20 to 30 of such archs; the present relief visible in the field is 2 to 5 meters. The type area of these landforms is Aaron Castellanos, in south Santa Fe (34°10' lat. S-62°30' long. W) (Figure 6).

Figure 6. Arcuate megadunes at Aaron Castellanos.

5.2. Grain Size of the Sediments - The Sand/Silt Limit in Wind Transport

The superficial sediments in the region are very fine silty sand and sandy coarse silt, with a media value about 4 φ (frequently slightly finer). Considering that the numerous analyzed profiles represent typical aeolian dunes, a rather uncommon feature turns up: many dunes are composed of 50 % of grains smaller than 62 μm. This is a point with a high theoretical interest, and leads to the question of the real (physical) lower limit of the fraction named "sand". Wentworth (1922) puts this limit in 62 μm, very probably based on the mathematical properties of his φ series. In the Soil Science the lower limit of sand is located in the value of 50 μm, which is as arbitrary as the former. An independent approach is provided by experiments in wind tunnels (Bagnold, 1965); there, the movement of individual sand grains produced by wind begins with diameter of 70 μm; below that diameter the viscosity of the air forms a semi-viscous film of laminar flow, which requires a larger wind force to deflate. Bagnold, based on the aerodynamical properties of the different grain diameters, proposed that the lower limit of sand should be 80 μm.

In the case of the PSS, and considering that dunes are sedimentary structures formed by transport of grains moved in creep and saltation, the comparatively small diameter of the individuals fits coherently with the classification usually applied in Germany, which includes the category of "powderish sand"(staubsand) between 62 and 20 μm. Coincidently, the Hopkins classification (at present in disuse) is a signal of attention, because in such classification the limit sand-silt is located in the diameter of 32 μm (Muller, 1967).

The grain size results (obtained by sieving for fractions > 37 μm) have not demonstrated to reach the necessary level of detail for discriminating the sedimentary processes occurring on the PSS. In consequence, grain-size analysis by Low Angle Laser Light Scattering (LALLS) were performed on cores from the Teodelina research borehole. 63 grain-size classes (0.05 - 750 μm) with intervals of 2 a 3 μm in the modal class were obtained for each sedimentary core (figure 11). Mean results indicate the following composition for both drilled Late Pleistocene formations (Kröhling, 2010): Carcarañá Fm: sand: 50%, silt: 43.26%, clay: 6.75%. Truncated soil by erosion on top of the Carcarañá Fm: silt: 70.15%, sand: 14.91 % and clay: 14.94% (with an enrichment of clay and a decrease of sands). Lower sector of Teodelina Fm: 63.08% silt, 25.55% sand and 11.38% clay. Middle section of the Teodelina Fm: 55.48% silt, 33.87% sand and 10.66% clay. Upper section of the Teodelina Fm: 66.70 % silt, 20.03% sand and 13.27% clay. The figure 11 shows for the Carcaraña Fm a unimodal distribution (Mo: 76-89 μm) with marked positive skewness (Mz and Md ≅63 μm). Teodelina Fm also presents a unimodal distribution (Mo: 56 - 65 μm for the lower part and Mo: 48- 56 μm for the upper section) and positive skewness (Mz>Md, in the coarse/medium silt).

5.3. Salt Weathering

The process known as "salt weathering"(Ollier, 1969) is an important mechanism in many places of the PSS. When rains occur, interdune depressions and other low sectors are transformed in temporary shallow lakes, filled with dirty water containing suspended clay and silt. Once water evaporates, a dry compact layer of fine sediments cover the depression. Such a layer, is not deflated by the wind itself, but it is destroyed by the salt crystals that grow with strong cristalization pressure inside the pores of the sediment. The result is a loose powder composed of sand-sized grains which are locally deflated to the margin of the depression, forming accumulations named "lunettes" or "clay dunes". Hundreds of interdune and other depressions are now transformed in temporary lakes, most of them bordered by lunettes at the leeside. Paleowind directions have been reconstructed on basis of lunette counting (Iriondo and Kröhling, 1995).

5.4. Dune Dissipation

This process was described by Bigarella (1979) for events observed in the coast of Brazil: large parts of dunes or entire dunes may undergo subsequent structural changes, causing disappearance of the aeolian cross-strata pattern. The original structures are replaced by much less clearly defined patterns or by structureless deposits, named dissipation structures. The new internal structure is characterized by an irregular, wavy and lenticular pattern accented by differences in concentrations of colloidal material (clays, hydrous iron oxides and humic compounds) (Figure 7).

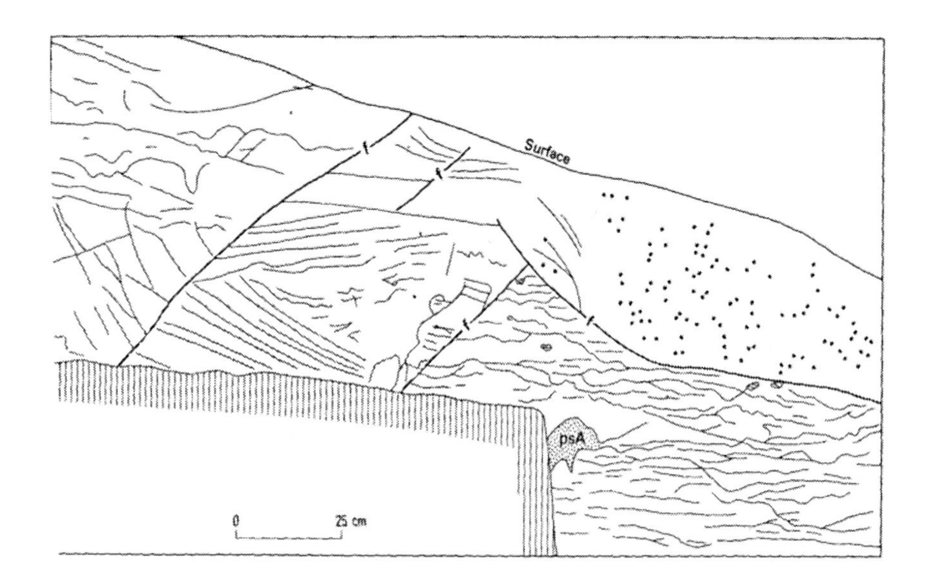

Figure 7. Dissipation structures (after Bigarella, 1979).

The mechanism of dune dissipation played an important role in the evolution of the landscape of the sandy Pampa. Dunes are particularly sensitive to pluvial erosion: each drop of water hitting the loose sand during a rain produces a small crater, dispersing the grains which are moved mostly downslope on the dune surface. Such a mechanism is enhanced by the rapid saturation of the superficial millimeters of the sand, which collapse in small transient contorted filaments and centimeter-sized "humid avalanches". This process is typical of semiarid climates, characterized by scarce vegetation cover, and indicates a climatic change from desertic to more humid representing periods with sporadic but intense rainfalls. The mechanism of dissipation flatted the Pampa landscape at least in two periods, at OIS3 and during the Holocene Optimum Climaticum (Iriondo and Kröhling, 1995).

According to the LALLS data of the cores of both formations of the Teodelina borehole, the deduced mechanisms of pure and modified saltation and the short-term suspension (according to the analysis proposed by Nickling, 1994) are responsible for the sedimentological characteristics of the dunes (Kröhling, 2010). The dissipation concept clearly explains the predominance of the deduced suspended mechanisms of transport. Obtained data are consistent with geomorphological and stratigraphic conclusions indicating that the incorporation of fines during the dissipation processes altered the initial composition of dunes.

5.5. Pedogenesis/Multisols

The PSS underwent several climatic changes in the upper Quaternary. During the humid periods, pedogenesis occurred in the region; paleosols and buried soil profiles of different ages are widerspread and help to reconstruct the climatic sequence. The most important humid period was the Optimum Climaticum of the Holocene (8.5-3.5 ka. BP in North Pampa); a multiproxi study revealed that during that time occurred intense pedogenesis down to 40° of latitude under a climate warmer than the present one and annual precipitation above 2,000 mm (Carignano, 1997; Iriondo et al., 2009) (Figure 8).

An interesting feature showed up during detailed mapping in the Pampa and surrounding regions: soil or paleosol bodies that, maintaining a identifiable continuity, bifurcates in two or more layers located at different levels of the sedimentary column. The layers coalesce, disappear or bifurcate again over distances of tens or hundreds of meters. Such structures are named "multisols" (Iriondo, 2009) and represent local (or regional) environments where pedogenesis and sediment accumulation or erosion co-exist.

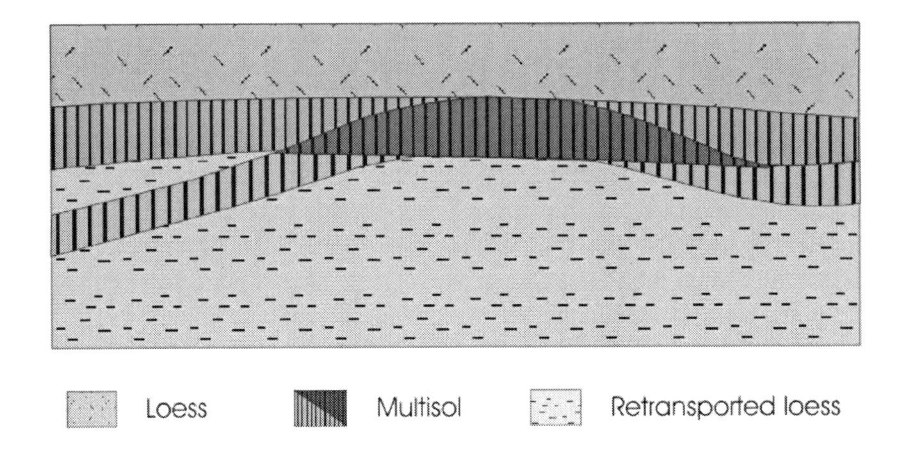

Figure 8. Multisol profile (after Carignano, 1997).

5.6. Carbonate and Iron Mobilization

According general knowledge (Ollier, 1969; Catt, 1991; Strahler, 1997), processes of lixiviation and elluviation of rocks and sediments occur under several types of humid climates, depending of the amount of annual precipitation and temperature. Tropical climates above 20° mean annual temperature and precipitation of 1000 mm/yr or more produce generalized movement of iron hydroxides and reddening of sediments, often with formation of iron concretions (Iriondo and Kröhling, 1997). Semiarid climates with precipitation between 300 and 700 mm/yr are characterized by ustic pedogenesis, which produces carbonate precipitates in the sediment profiles. Below 300 mm/yr (arid climates) precipitation of gypsum occur in playas and phreatic levels. These indicators resulted useful in multi-proxi analyses of paleoclimates of the region (Iriondo et al., 2009).

5.7. Hydrographic Characteristics

During humid periods (as the present climate) the Pampa is subject to an environment characterized by water movements and not by wind dynamics. In such cases, infiltration dominates in a landscape formed by flat surfaces and local closed depressions. Morphological characteristics of the PSS are associated with the blockage of the preexisting and marginal drainage by the dunes. Under those conditions, many depressions are transformed in shallow lakes feed by seepage and dissipation of dunes is enhanced and begins the formation of hydrographic nets under rather particular conditions. Water divides often are not-well-defined flat surfaces. Drainage networks begin in a kind of non-permanent wetlands, which move very slowly downwards as "mobile marshes" (Iriondo and Drago, 2004). The water masses eventually occupy dells (shallow, wide long depressions of tectonic origin or old fluvial channels), where the flow is measured in a few meters per day. Finally the water enters in normal channels.

6. Sequence of Events

Regional studies made in the last decades, including luminescence dating established a sequence of events occurred during the Late Quaternary in the Pampa and neighboring regions (Iriondo and García, 1993; Iriondo, 1999; 2010). Widespread dune activity occurred on the PSS during the OIS4 and OIS2 (mainly LGM). Subsequent periods of aeolian activity have reworked older deposits, mainly during the Late Holocene. In short, the environmental sequence was the following (Figure 9):

Isotopic State 4 - According to the available information, the PSS was accumulated during the OIS 4 (roughly from 77 to 65 ka. BP; Iriondo and Kröhling, 1995). The OIS4 appears in the South American literature as the coldest period in the last glacial-interglacial cycle, with the maximal extension of glaciers and lowest snow-line in the continent (Clapperton, 1993). The origin of the PES is coherent with that scenario.

This phase of the PSS was characterized by the formation of major longitudinal megadunes with SSW-NNE and S-N directions, with a gentle anti-clockwise curvature. Such geoforms are at present only visible in satellite images, and hardly perceptible in aereal photographs, owing to the advanced dissipation occurred in later processes during sub-humid climates. The area originally covered by megadunes was approximately 100,000 km^2 at the center and west of the system. Each of those dunes were 50 to 200 kilometers long and 3 to 5 kilometers wide; they are grouped in regular sets.

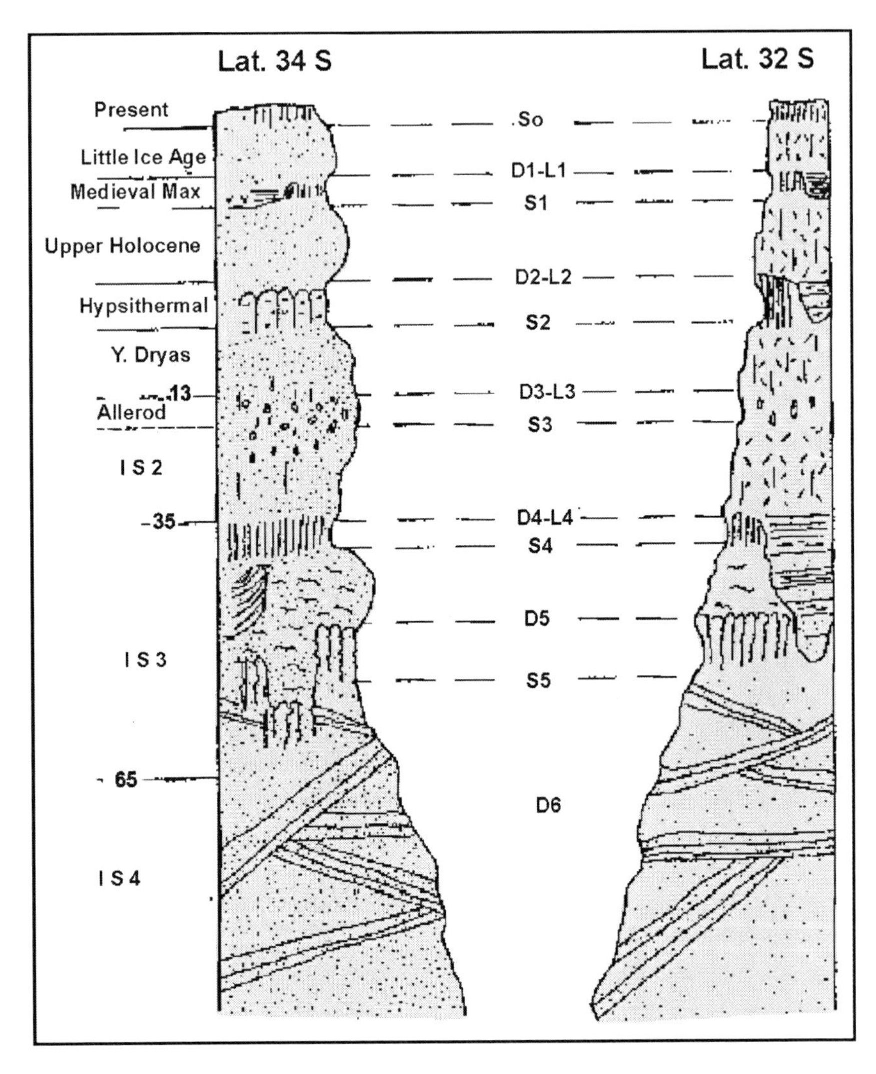

Figure 9. Sequence of climatic events in the Pampean Sand Sea during the upper Quaternary (after Iriondo and Kröhling, 1995).

According to the dimensions of similar dunes actives at present (Mc Kee, 1979; Short and Blair, 1986) the original elevation should have been between several tens to more than a hundred meters. Internal structures of megadunes outcrop in deflation depressions and in few roadcuts. The sedimentary profiles are composed of large sets, several hundred meters long and up to five meter thick, dipping 10 to 20°.

Other types of megadunes were probably formed in the east of the PSS during the OIS4; in south Santa Fe some sand accumulations have diameters of several kilometers and up to 20 meters in thickness, suggesting original a star type of megadune.

Isotopic State 3 – The OIS3 (65-36 ka. BP) is characterized in the PES and neighboring regions by a climatic improvement with irregular oscillations. The climatic change from OIS4 was probably rapid, because the sediments of this period lay concordantly on the aeolian sands, without any stratigraphic discordance.

Basically, the OIS3 was formed in the region by a sequence of three climatic phases: a) A humid phase that generated a soil at the top of the dunes; such a soil level reproduces the dune relief. b) A period of generalized dissipation of dunes that produced the flattening of the inherited geoforms, which is interpreted as a semiarid climate. c) A second humid phase that generated a second soil level.

Important fluvial belts were formed during the OIS3 in the sand sea, some of them were larger than the present ones. A dating of 45,610 ± 1990 yr. BP was obtained in one of such belts.

Latrubesse and Ramonell (2010) reported OSL dates of the longitudinal megadunes in Buenos Aires and San Luis provinces with ages of 43 ka. AP and 41 ka. AP, respectively. We interpreted those results as the age of the dissipation of the dunes occurred mainly during the OIS3.

Isotopic State 2 – The OIS2 was marked by a lowering in temperature in the whole planet, which provoked a generalized advance of glaciers in the Andes Cordillera. The Patagonian climate migrated to the NE, resulting in a dry and cold environment in the Pampa (Iriondo and García, 1993; Iriondo, 1999). According to several OSL dates, it e began around 36 ka. BP and finished at 8.5 ka. BP (in the Early Holocene) in the Pampa region. A major period of aeolian sand movement occurred during the latter part of the last glacial period. In several areas of the PSS were generated dune fields that advanced to the north and covered the fluvial net developed in the OIS3. An important mobilization of atmospheric dust accumulated in the leeward position (to the north) and formed a wide belt of loess (Zárate, 2003; Kemp et al., 2004; Iriondo and Kröhling, 2007, among others).

OSL ages of Tripaldi y Forman (2007) on samples taken from the San Luis dune field (northwestern part of the PSS; the Blowout Section) indicated ages of 33 and 28 ka. BP for both oldest aeolian units representative of the LGM. Latrubesse and Ramonell (2010) cited OSL datings for a dune field of

Buenos Aires province (central part of the PSS) with results close to 30 ka. BP.

Termination of the Pleistocene - During the last thousands of years of the Pleistocene and first times of Holocene (approximately from 14 ka. BP to 8.5 ka BP) the climatic dynamics underwent a change, the dominant winds were the Westerlies up to the latitude of 33° South. Such scenario was dominantly erosive, resulting in hundreds of aeolian depressions.

A discontinuous sedimentation of sandy loess occurred together with incipient pedogenesis, producing a particular complex of soil horizons named "multisols" (Iriondo, 2009). The dissipation and destruction of dunes occurred in the eastern half of the sand sea. This climate was roughly simultaneous with the European Younger Dryas. Latrubesse and Ramonell (2010) indicated an OSL age of 9 ka BP for small longitudinal dunes in the southern of San Luis province (PSS).

The Holocene *Optimum Climaticum*- It was a general warming of the climate on Earth occurred around middle Holocene (8.5 to 3.5 ky BP in the North Pampa). Such a warming produced an increment of precipitations, resulting in a soil-forming climatic period, with very low incidence of the wind down to the north of Patagonia. Fluvial belts and fluvial nets were reconstructed in this time.

Late Holocene – During the late Holocene occurred a dry climatic pulse, produced by a blocking anticyclone that developed on the Pampa and neighboring regions. It remobilized large volumes of sand in the eastern sector of the PSS, forming a field of megadunes of parabolic type in south Santa Fe and southeast Córdoba provinces (with OSL ages between 3 to 1.4 ka. BP; Iriondo and Kröhling, 1995; Kröhling, 1999a,b; Carignano, 1999).

The general direction of the wind was north-south. In the west of the system (San Luis province) turbulent irregular winds, weaker than those of the OIS2, flowed in SSE-NNW direction, with anti-clockwise curvature (Ramonell et al., 1992). To the north, the buried soil developed on the loess belt was partially deflated and re-sedimented as a thin loessic formation by anticyclonic circulation (Iriondo, 1990).

The Medieval Maximum – Several indicators of a warm and humid climate occurred between 1,4 and 0,8 ka. BP are detected in the Pampa region. Such indicators are pedological, geological, archaeological and faunistical. In the west of the PSS small and thin deposits of green loam inside deflation depressions indicate the existence of permanent shallow lakes around AD 1,000. Such depressions are at present completely dry and subject to deflation.

7. The Eastern Region of the Pampean Sand Sea

The eastern region of the PSS is located in the more humid sector of the Argentine plain, therefore the arid periods are weaker and the humid events are more marked than in the western sand sea, that has 900 km in east-west direction.

7.1. Local Area Teodelina

Teodelina represents an area of some 10,000 km^2 located at the northeast belt of the PSS in the province of Santa Fe and Buenos Aires. The morphology is characterized by extensive flat surfaces (top of the Teodelina Fm), which included minor hilly areas formed by the dune fields formed by the San Gregorio Fm. Four study boreholes were made in the area, being the more important drilled in Teodelina (34°11 lat.S, 61°31′ long.W, 88 m asl; 15 m drilled) and in San Gregorio (34°17′lat.S and 61°55′long.W; 102 m. asl; 5 m drilled).

Quaternary Stratigraphy

Carcarañá Formation – The oldest geological unit in the area is the *Carcarañá Formation*, produced by dissipation of aeolian dunes (Kröhling, 1999). The upper section outcrops in quarries and was perforated in the boreholes. It is formed by very fine sand to silty very fine sand with scarce clay, reddish brown (5YR 5/6) in humid and yellowish brown (10YR 3/6) in dry. It is organized in poorly defined horizontal very thick strata, with concordant contacts, without visible internal structures. With moderate degree of compaction and weak structuration in resistant, fine to medium blocks. The sediment is non-calcareous. Local bioturbations (root molds and crotovines) appear in isolated spots. The lower contact of the unit was not reached in the subsoil; the upper contact is marked by an erosive discordance, below Teodelina Formation. Grain size analyses indicate a silty sand composition, with bimodal distribution (250125 μm and 88-74 μm). The main aeolian facies includes locally dissipation structures, characterized by an undulated pattern, continuous, several centimeters thick, marked by different concentrations of fine and colloidal materials inside the sand (Bigarella, 1979; Kröhling, 1999).

The Carcarañá Fm is composed of volcaniclastic materials originated in the Andes (acid volcanic glass, policrystaline quartz, subordinate plagioclases and lithic fragments) with scarce participation of sediments coming from the metamorphic Pampean Ranges (monocrystaline quartz and plagioclases). Heavy minerals contribute with less of 3 % of the total mass (pyroxenes, hornblende, tourmaline, micas and magnetite, among other). The abundant vitroclasts found in Teodelina compose four populations: a) Colourless shreds with fluidal textures, partially devitrified or with high birefringence along microchannels and vacuoles; b) Deeply altered laminae, generally altered to sericite; c) Colourless blocks with smooth surfaces and incipient devitrification in the central part; d) Triangular- to tubular shreds, partially devitrified. Felspars are also grouped in four populations.

Relicts of an intraformational paleosol were found in the Carcarañá type area (Kröhling, 1999). The age of this unit corresponds to the OIS 3; a TL dating indicates an age of 52,310 ± 1,200 yr. BP.

Paleosurface – A complex stratigraphic level composed of fine sediments, carbonate segregations, and a paleosol developed at the top of the Carcarañá Fm indicate the existence of a stable surface generated in a humid event occurred during the OIS3. Such paleosol outcrops as relicts several meters long that appear at variable positions in the profiles. It is reddish brown in color and is moderately structured in very firm, fine blocks. Root molds with internal black films and rizoconcretions are common.

A carbonate precipitate was formed in local depressions, forming a hard rock characterized by a net of thin veins with dominant horizontal extension, a structure that indicates a phreatic origin for this feature. The typical thickness varies from 40 to 80 cm. This layer is structured in subangular blocks 2 to 3 cm long. The general color is brown, with olive green sectors which are probably composed of altered volcanic ash.

The bottom of the collector of the present fluvial net is a part of the paleosurface. It is formed by a sandy clay, brown in color. It is consistent to hard and includes abundant carbonate precipitates, from small concretions and veins to blocks several square meters large. In those cases, the deposit is a right sandstone with carbonate cement, composed of very fine sand grains and epigenetic millimetric micaceous laminae. The cementation produced a complex of 40 to 50 cm wide vertical polygons. A second generation of precipitates, represented by iron and manganese minerals, fills up small pores. This level contains Pleistocene fossils (Lujanense mammal fauna).

Teodelina Formation – This formation is the most important stratigraphic unit in the landscape of the eastern PSS. It was sedimented as dune fields during the main phase of the Last Glacial Maximum, subsequently dissipated in one or more sub-humid events, resulting now in extensive flat surfaces. It correlates laterally with the Pampean Tezanos Pinto loess (Kröhling, 1999; Iriondo and Kröhling, 2007b; Kröhling et al., 2010).

The Teodelina Fm is a very fine to fine silty sand to sandy silt, strong brown in humid (7,5YR 4/6) to yellowish brown in humid (10YR 6/4), friable, non calcareous. The grain size analysis (obtained by sieving) reveals the existence of three populations of grains, which were produced by different mechanisms of aeolian transport. The principal and secondary modes are located in the fractions 250-125 μm and 63-53 μm. Fractions lower than 37 μm have a considerable participation in the total mass of the sediment (41 % in average). It is remarkable the lack of intermediate fractions in all the sediment. The registered fractions correspond to three types of mechanisms proposed by Nickling (1994): pure saltation, short suspension and long suspension. We believe that the first two mechanisms were produced in the original dune fields and the materials transported by long suspension were incorporated during the subsequent dissipation phase.

The mineral composition of the modal fraction 53-62 μm is dominated by shreds of volcanic glass (64.1 %), with low participation of feldspars (18.1), scarce quartz (6.6 %) and alterites (6.6 %) and others. The percentage of heavy minerals reaches the 3.7 %, most of them amphiboles and pyroxenes. Roundness and sphericity vary from low to moderate, with some cases of high sphericity.

Two populations of different sources were detected: the most important one has volcanic origin, composed of volcanic glass with accessory albite and quartz, with S (sphericity) between 0.2 and 0.4 and R (roundness) from 0.3 and 0.5; the second population derives from the Pampean Ranges, it is formed by quartz, potasic feldspars, microagglomerates and lithics (S from 0.6 to 0.9 and R from 0.3 to 0.8). Superficial textures also show two populations.

The vitroclasts, as dominant component, were studied in more detail. The degree of alteration and attrition allowed the following discrimination: 1) Fresh volcanic glass: a) triangular and rectangular shreds, with curly fringes and marked vertices, also tubes and vacuoles (16.3 %); b) colorless plates, with very scarce alteration at the centers (0.6 %). 2) Partially altered glass: a) angular colorless shreds with fluidal textures affected by devitrification processes, showing moderate to high birefringence along channels and vacuoles (27.6 %); b) subrounded fragments, with moderate to high sphericity, most of them devitrified (5.6 %); c) altered shreds, some of them including microlites and green micro-crystals (14 %).

The coarser mode (250-125 μm) has a clearly different mineralogical composition, as same as the fraction 88-74 μm; there, quartz dominates with proportions of feldspars (acid plagioclases), and muscovite and calcite as minor components. Cristobalite was detected in the 52-62 μm fraction.

Final Pleistocene loess – That deposit is a sandy loess, 1 to 2 meter thick, accumulated at the end of the Pleistocene and Early Holocene. Yellowish brown in color (10 YR 5/4), it forms columnar profiles, partially affected by columnar disjunction. The compaction is slightly larger than the low-lying Teodelina Fm. The unit is capped by a moderate pedogenesis at the top.

Grain size analysis showed that the sediment is a fine to very silty sand (56 % coarser than 62 μm) with the principal mode in the 74-88 μm fraction, and a second mode in the 53-62 μm mode. This unit correlates with the upper section of Tezanos Pinto Fm (Kröhling, 1999; Kröhling et al., 2010), sedimented at the late Pleistocene/lower Holocene (14 to 8 ka. BP).

This deposit is discontinuous; it forms isolated spots in the region (Figure 10 and 11).

San Gregorio Formation – San Gregorio Fm is the stratigraphic name of the megadune fields located in the Teodelina and Canals areas, and other similar and coetaneous dune fields in the Pampa. It is composed of loose fine to very fine sand, yellowish brown in color. From a geomorphological point of view, such megadune fields constitute the relief superimposed to the flat level of Teodelina Fm and isolated spots of loess (Iriondo and Kröhling, 2007b).

A short description of the type profile of the formation is the following (from base to top):

0.00-4.80 m – Very fine to fine sand, loose, yellowish brown, massive.

4.80-7.30 m – Very fine to fine sand, yellowish brown in color, massive in general, including sectors with diffuse lamination.

7.30-8.30 m – Sand similar to the former. Grayish brown in color owing to an incipient pedogenesis. Slightly more consolidated.

Grain size analyses made on sediments of the Teodelina area indicate very fine to fine sand, with variable content of silt (from 37 % at the base to 4 % at the top). The mode is located in the 74-88 μm fraction, with most of the sediment mass between 125 and 74 μm. The mineral composition of the modal fraction is formed by glass shreds (38.4 %), alterites (26.1 %) and feldspars (21.8 %), with quartz as minor component (9.3 %) and others.

The percentage of heavy minerals is relatively high (3.8 %) with dominance of metamorphic minerals (garnet, hornblende, etc.) originated in the Pampean Ranges (Figure 12).

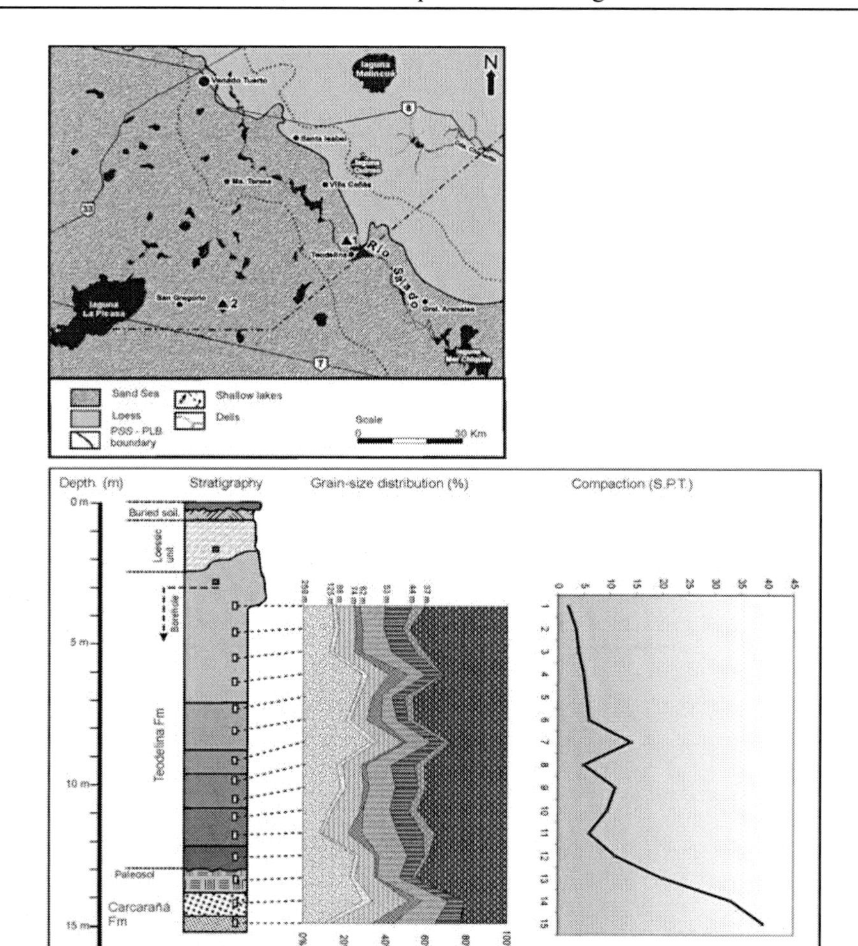

Figure 10. Teodelina map and profiles (after Iriondo and Kröhling, 2007).

The whole volcanic glass grains appear in several different populations:

a) Colorless subangular fresh plates, some of them including microcrystals (5.2 %).
b) Colorless plates devitrified at the central area (2.8 %).
c) Colorless shreds partially devitrified along channels and vacuoles (17.8 %).
d) Totally altered shreds, subrounded to rounded (11.8 %).
e) Brown color glass (0.8 %). Felspars were divided in four classes, quartz in two.

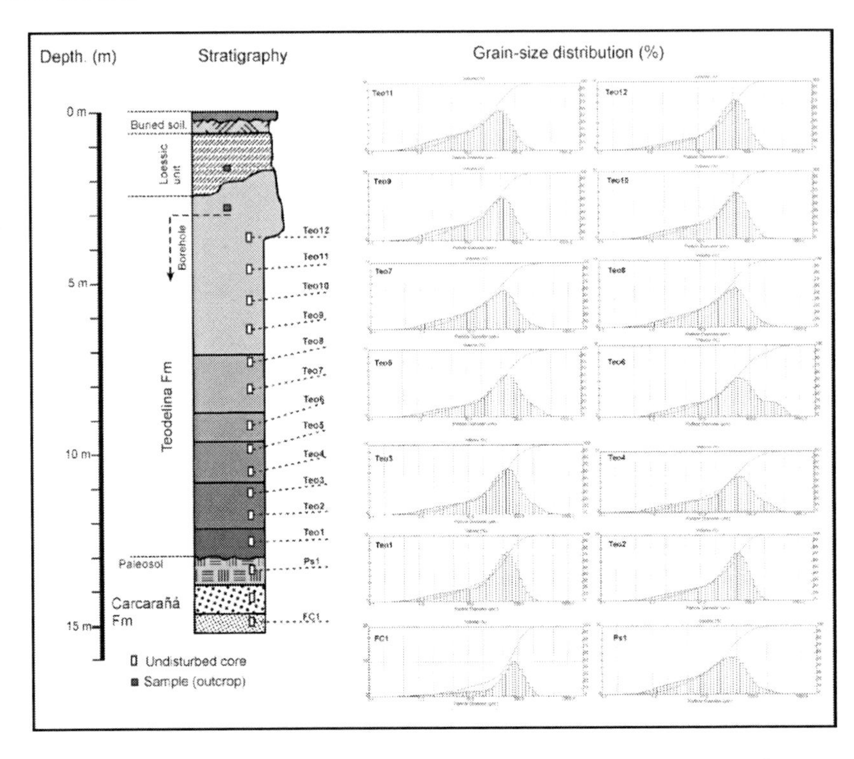

Figure 11. Grain size distributions in the stratigraphic profile of Teodelina (after Iriondo and Kröhling, 2007).

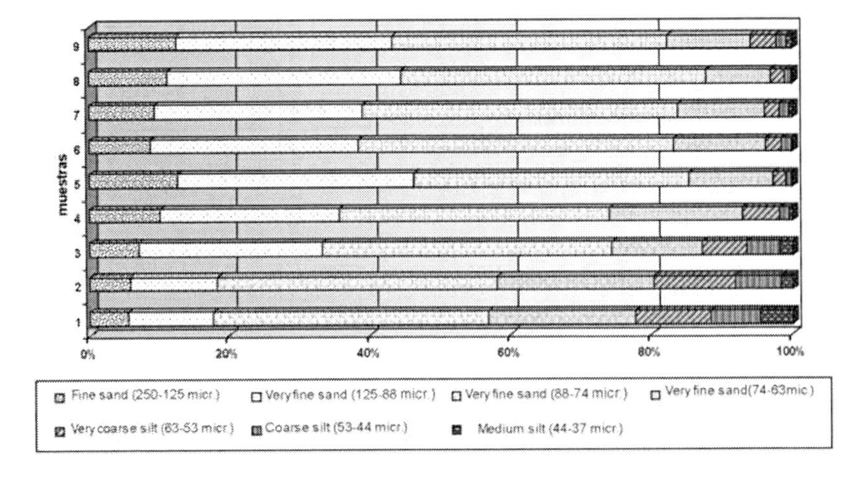

Figure 12. Grain size distribution of the San Gregorio Fm in the type profile (after Iriondo and Kröhling, 2007).

Most quartz and feldspar grains have high sphericity and intermediate roundness. The mineralogical composition of this formation is homogeneous along large distances, at least 70 kilometers to the SE (General Arenales) and 250 km to the NW (Canals).

7.2. Local Area Canals

Canals area represents 7,600 km^2, with a landscape characterized by a field of fish-scale and parabolic megadunes (San Gregorio Fm) cover discontinuously a sandy formation (Tedelina Fm, with the top forming extensive horizontal surfaces in the region). Twelve study boreholes were made in the area, being the more important drilled in Canals (33°32′lat.S and 62°57′long. W; 124 m asl, 26 m drilled) and in San Gregorio (34°17′lat.S and 61°55′long.W; 102 m. asl; 5 m drilled) (Figure 13).

Figure 3. A general satellite view of Canals (contact between fish-scale and parabolic fields).

Stratigraphy

Paleosurface – In Canals this unit is composed of a consolidated brown sand, very fine, with 6 to 7 % coarse silt. The grain size mode is in the 74-88 % fraction and composes 49 to 53 % of the sediment mass. In La Cesira (in the same area, 50 km to the south) this unit outcrops in a quarry. The visible thickness is there 3 m. The material is arenaceous to silty arenaceous, consolidated to compact, clear yellowish brown in color (10YR 6/4) with variations to the light olive containing volcanic ash. Carbonate segregations reach 40 % of the total mass in some spots, the structures of such mineral have horizontal irregular, undulate structures, suggesting a phreatic origin for the precipitate. The sediment is structured in firm subangular blocks. Rootmolds and micropores with films of iron and manganese minerals. In erosive discordance lays the Teodelina Fm.

Teodelina Formation – Very fine silty sand, moderately compacted, yellowish brown in color (10YR 5/4), including carbonate and gypsum concretions. The grain size mode is located in the fraction 74-88 μm. This deposit, with a thickness of 5 to 7 m in the area, constitutes the horizontal surface of interdune flats. Pleistocene megafossils (Glyptodon sp.) were recovered from excavations. In La Cesira quarry this deposit lays in discordance on the paleosurface as a friable silty sand, brown in color (10YR 5/3), including fragments of the older deposit at the base. Such material has gravel size, well rounded and with scarce selection; the upper part of the deposit is a dissipation sandy facies, ordered in irregular undulose small bodies. Internal channel infills, up to 50 cm thick and 1-2 m wide, composed of dark brown loam (10YR 3/2) are included in the profile.

San Gregorio Formation – This unit forms the body of the field of megadunes covering 7600 km^2 in the southwest of the Córdoba province (33°30' to 34° lat.S and 62° to 63° long.W and surroundings). It is formed by loose very fine sand with moderate to scarce coarse silt; the mode is in the fraction 74-88 μm and the color is light yellowish brown (10YR 6/4). The deposit is formed by very coarse strata 60 to 110 cm thick, with internal lamination or very fine stratification (2-4 cm thick). In artificial cuts at roads, this formation produces subvertical slopes with subfussion holes and tubes.

Little Ice Age superficial sand – The sand deposit covering the megadunes of San Gregorio Fm contains rests of European mammals (Bos taurus and Equus) and also European artifacts, such as bricks and metal pieces. According to colonial references (Parras, 1943) the region was a desert in the XVII Century. The wind action produced a general reworking of the surface of the dunes, down to one to two meter deep. In the Canals area dominated the accumulation on the dunes and deflation in the interdune flats and in the central depressions of dunes.

Irregular accumulations of sand up to 200 m long and 2 to 4 m deep cover that part of the landscape. The sand is very fine, monomodal (mode in the 88-74 μm fraction), brown in color (10YR 5/3). The deposits are formed by strata 20-25 cm thick, subhorizontal to horizontal, with internal very fine planar stratification (2-4 cm) with corrugated lines in some cases (those indicate dissipation processes). In some flanks of megadunes the strata are thicker (40 to 60 cm) following the convex original surface of the terrain; such strata have diffuse internal lamination and millimetric nodules of biological origin, formed by silt particles (Figure 14).

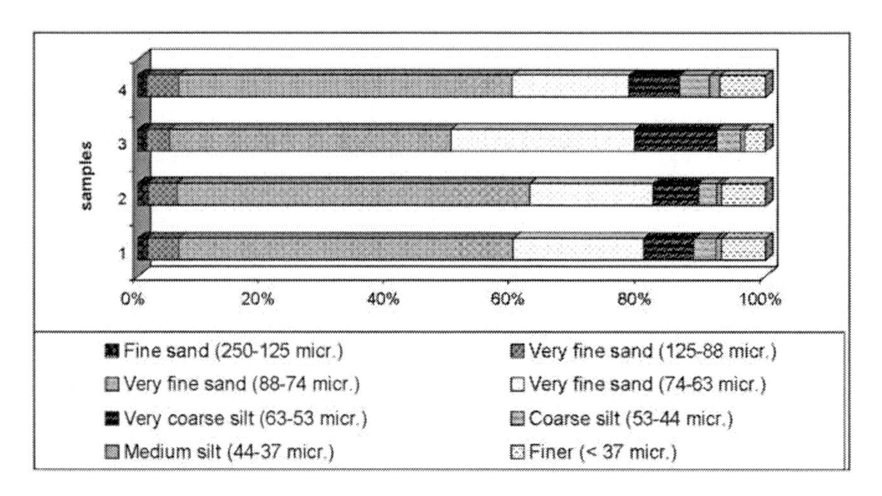

Figure 14. Grain size distributions of the geological units outcropping in the Canals area – 1: Paleosurface; 2: Teodelina Fm; 3: San Gregorio Fm; 4: Little Ice Age superficial sand.

7.3. Transition with the Peripheric Loess Belt

The leeward limit of the sand see marks the line where the sand, conveyed by the wind in saltation and creep, is replaced by the loess, formed by silt and clay transported in suspension by the wind and trapped by the vegetation on the surface of the terrain. That limit is several hundred kilometers long in the Pampean Aeolian System; it is a particular zone 5 to 15 km wide, characterized by overlaps of sand and silt, spots of loess with flat surface included in sandy hills and areas of sand inside a loessic landscape. Same of such areas cover dozens of square kilometers, the largest one in the northeast of the system is the Hansen Sand Field in the province of Santa Fe.

The PPS/PLB transition in north-south direction was analised between the localities of Arias and Cavanagh, at the longitude of 62°15'W, in the province of Córdoba. In a distance of 20 km, the southern extreme (PSS) is characterized by large megadune fronts with a superimposed rugosity formed by small hills 2 to 4 m high in a chaotic pattern and depressions transformed in ponds.

The sediment is very fine sand containing regular proportion of coarse silt, with mode in the 74-88 μm fraction and strong positive skewness. 10 km to the north the sand is covered by a 25 cm thick layer of the San Guillermo Fm, a grey sandy silt which belongs to the PLB. 10 km farther to the north (at Cavanagh) the terrain is composed of loess (Tezanos Pinto Fm) which form most of the PLB with a flat and relatively elevated surface including isolated low hills; round deflation hollows about 200 m in diameter (transformed now in shallow ponds) are scattered in the landscape. The sediment is a bimodal loam, with modes at 44-53 μm and 62-74 μm.

8. Analogies with the Present Climate

This approach for the reconstruction of past environmental conditions in the Pampa is based on the theoretical principle that the atmosphere is a thermodynamic machine with only few degrees of freedom: when a part of the system (for example, the Pampa or the Amazonia) receives a specific influence through the border conditions, the internal response would be always the same. That is the robust axiom of the weather forecast, and can be advantageously applied in reconstruction of the broad meteorological parameters of dominant scenarios of the past (Iriondo and García, 1993; Iriondo et al., 2009).

Related disciplines, as Palinology and Paleobotany, use customarily this kind of theoretical basis, reconstructing past climates based on the ecology of living plants. In short, the basic characteristic of climate at regional scale is the repetition of synoptic meteorological structures along successive climatic periods in the same region.

An example of an structure is the already mentioned "Pampero" wind, a cold and dry SSW wind that occasionally blows in the Pampa transporting Pacific ocean air masses, and was the dominant, dune-forming synoptic structure during the OIS4. Another example is the anticyclonic circulation over the Pampas in extremely dry years, a phenomenon dominant in the late Holocene (Iriondo, 1990) (Figure 15).

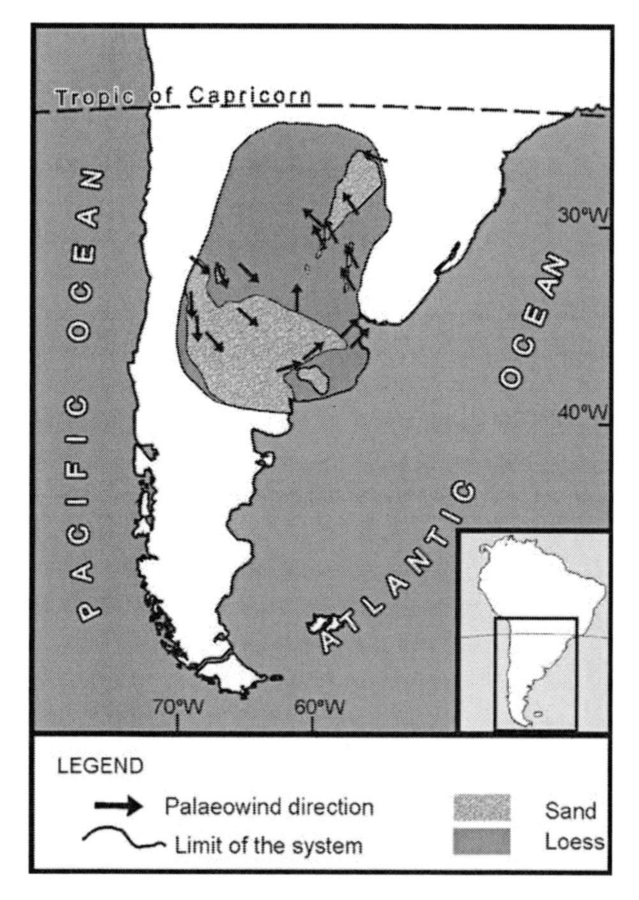

Figure 15. Dominant circulation of winds and sediments deposed at the Late Holocene during the accumulation of San Gregorio Fm (Iriondo, 1990).

Dunes are sensitive to modifications in the atmospheric parameters, such as wind direction and intensity and changes in precipitation. Conversely, pedogenic processes modifying wind-blown sand or dust, indicate that significant changes in climate did occur and paleosols are important proxies for marking precipitation and temperature.

Hence, some parameters remain directly registered in the geological and geomorphological archive. Moreover, it can be supposed with reasonable certainty that climatic parameters not preserved in the geological and geomorphological record have also occurred in coherency with the preserved ones. Some of those parameters, such as relative air humidity, frost frequency, maximal temperatures, etc., are undoubtedly important in the reconstruction of past environments. Hence, by applying this approach, the dominant synoptic meteorological structures of two contrasting climates of the Holocene of the Pampa were reconstructed by comparison with the extreme years of the XX Century (Iriondo et al., 2009) (Figure 16).

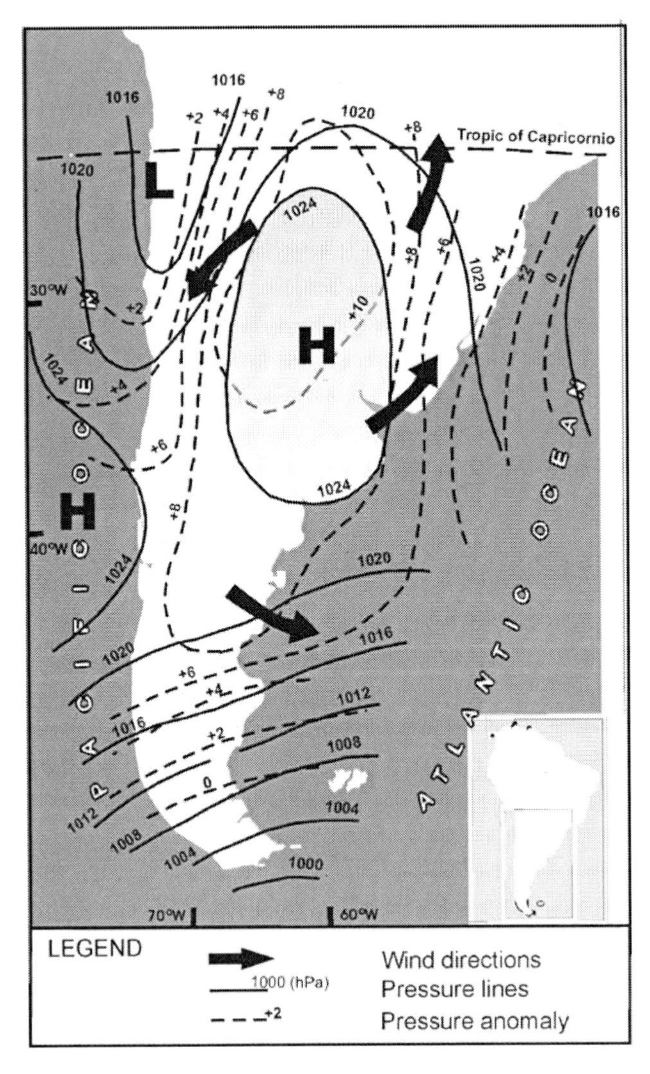

Figure 16. Anticyclonic circulation during an exceptionally dry year (year 1962) (after Iriondo et al., 2009).

The Holocene Climatic Optimum was a ca. 5000 year-long period, characterized by a humid climate in the Pampa. By applying the forementioned proxies, its typical scenario was:

- Mean annual precipitation ca. 2,200 mm.
- Mean annual temperatures around of 21°C, less than 1°C higher tan today.
- Mean maximum annual temperatures around 27°C.
- Mean minimum annual temperatures around 16°C, close to 1°C higher than today.
- Thermal amplitude smaller than today, virtually without frosts.

The Late Holocene period that generated the megadune fields gave the following values:

- Mean annual precipitation ca. 350 mm.
- Mean temperatures values of 15°C.
- Maximum annual temperatures of 22°C.
- Minimum annual temperatures of 8°C.
- Important shifts in minimum temperatures in the dry season.
- Thermal amplitude larger than today.
- Higher number of frosts per year, stronger winds, lesser relative air humidity and major evaporation rates if compared with today.

CONCLUSIONS

From the above information, the following conclusions can be mentioned (Figure 17):

- The Pampean Sand Sea covers a surface of 200,000 km^2, which is similar to other located in the great deserts of the World. -In contrast to the main tropical deserts, the bulk of the sediments was produced by glacial and nival processes. -Three main sources of materials are recognized: the Andes cordillera, the Pampean Ranges and volcanic explosive events (in that order). -A sequence of dry/cold and humid/warm climates produced a series of landscapes and dynamic modifications in the region. -The grain size composition of dunes could be a call of attention for the possibility of re-discussing the sand/silt limit. -Two of the dry intervals generated fields of megadunes: the OIS4 period (longitudinal) and the late Holocene (fish-scale, parabolic and arcuate).

– Humid intervals produced pedogenesis and a non-classical development of fluvial nets, characterized by undefined divisories and transient marshes as first-order conveyors of water.

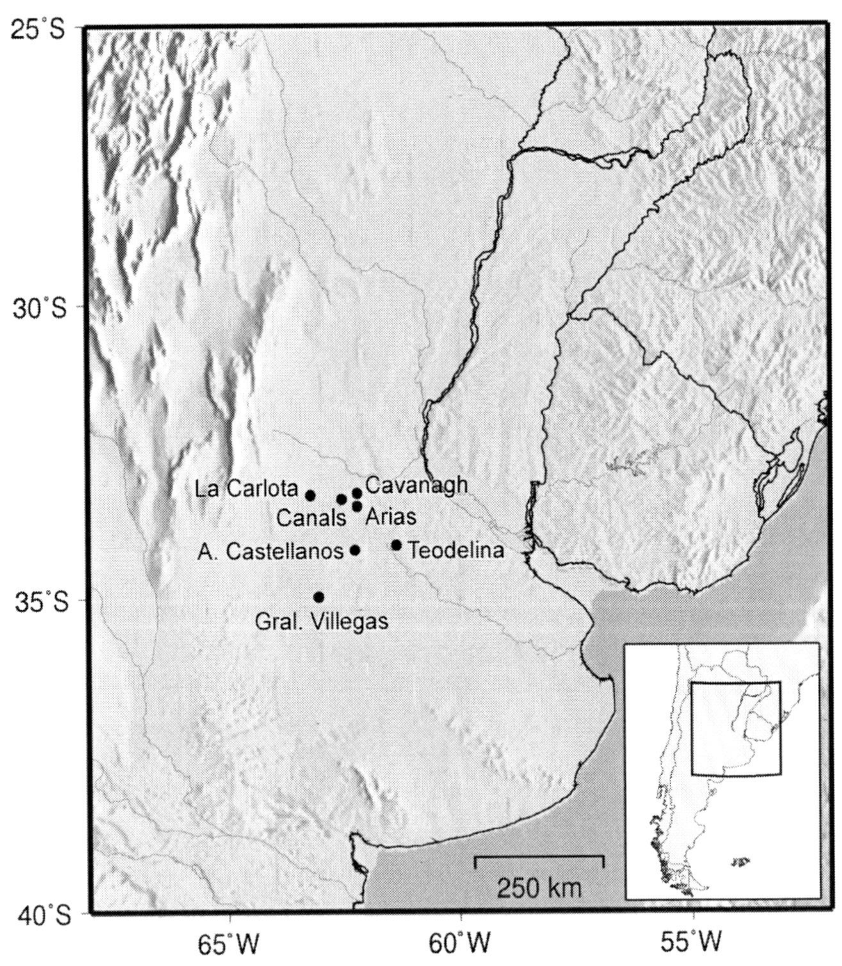

Figure 17. Location of the localities mentioned in the text.

REFERENCES

Ahumada, A. 1990. Ambientes, procesos y formas periglaciales o geocriogénicas en Quebrada Benjamín Matienzo, Cordillera Principal, Mendoza. *Revista de la Asociación Geológica Argentina*, 45(1-2):85-97. Buenos Aires.

Bagnold, R. 1965. *The Physics of blown sand and desert dunes*. Methuen and Co. Ltd. London, 265 pp.

Bigarella, J. 1979. Dissipation of dunes, Lagoa, Brazil. In: E.D. Mc Kee (Ed.): A Study of Global Sand Seas. *Geological Survey Professional Paper* 1052, Chapter E, 124-134, Washington.

Carignano, C., 1997. Caracterización y Evolución, durante el Cuaternario superior de los ambientes geomorfológicos extra serranos en el Noroeste de la Provincia de Córdoba. Doctoral Thesis. Universidad Nacional de Córdoba, 208 págs.

Carver, R. 1971. Procedures in Sedimentary Petrology. Wiley Intersciences, New York, 653 pp.

Clapperton, Ch. 1993. *Quaternary Geology and Geomorphology of South* America. Elsevier, 779 p. Amsterdam.

Goudie, A. S. , 2002. *Great Warm Deserts of the World*. Oxford. University Press, Oxford.

Goudie, A. S. 2006. Dune, Aeolian. Encyclopedia of Geomorphology (Ed: A.S. Goudie),volume 1; 1156 p.

GRASS Development Team, 2005. *Geographic Resources Analysis Support System* (GRASS), GNU General Public License. Eletronic document, http://grass.itc.it

Grohmann, C. 2005. Trend-surface analysis of morphometric parameters: A case study insoutheastern Brazil. Computers and Geosciences, 31:1007-1014. Elsevier.

Iriondo, M. 1987. Geomorfología y Cuaternario de la provincial de Santa Fe. D'Orbignyana,4:1-54. Corrientes (Argentina).

Iriondo, M. 1990. A late Holocene dry period in the Argentine plains. Quaternary of SouthAmerica and Antarctic Penninsula, 7:197-218. A.A.Balkema, Rotterdam.

Iriondo, M. 1992. Geomorphological map of the South American plains; 1:5,000,000 scale.Conicet/National Geographic Soc. Grant 4127/88.

Iriondo, M. 1997. Models of deposition of loess and loessoids in the upper Quaternary of South America. Journal of South American Earth Sciences, 10(1):71-79. Elsevier.

Iriondo, M. 1999. Climatic changes in the South American plains. Quaternary International,57/58:93-112. Pergamon Press.

Iriondo, M. 2009. Multisol: A proposal. Quaternary International, 209(1-2):131-141. PergamonPress.

Iriondo, M. 2010. La Geología del Cuaternario en la Argentina. Museo Provincial de CienciasNaturales "Florentino Ameghino", 437 pp. Santa Fe (Argentina).

Iriondo, M. and Brunetto, E. 2008 . El MAP en el sureste de Córdoba. 17° Congreso Geológico Argentino (AGA), Abstracts, pp. 1224-1225 , San Salvador de Jujuy. Argentina.

Iriondo, M. and Drago E. 2004. The headwater hydrographic characteristics of large plains: thePampa case. Ecohydrology and Hydrobiology, 4(1):7-16.

Iriondo, M. and García, N. 1993. Climatic variations in the Argentine plains during the last 18,000 years. Palaeogeography, Palaeoclimatology, Palaeoecology, 101:209-220.

Iriondo, M. and Kröhling, 1997. The tropical loess. Proceedings of the 30th. International Geological Congress, 21:61-77. VSP. Beijing.

Iriondo, M. and Kröhling, D. 1995. El Sistema Eólico Pampeano. Contribuciones, Museo Provincial de Ciencias Naturales Florentino Ameghino, 5(1):1-68. Santa Fe.

Iriondo, M. and Kröhling, D., 2007. Non-Classical Types of Loess. In: B.Flemming y D. Hartmann (Eds.), From particle size to sediment dynamics (Special Issue). *Sedimentary Geology*, 202 (3), 352-368

Iriondo, M., Brunetto, E. and Kröhling, D. 2009. Historical climatic extremes as indicators for typical scenarios of Holocene climatic periods in the Pampean plain. *Palaeogeography, Palaeoclimatology, Palaeoecology*, 283:107-119. Elsevier.

Jarvis, A., Reuter, H., Nelson, A. and Guevara, E., 2008. Hole-filled seamless SRTM data V4, International Centre for Tropical Agriculture (CIAT), available from http://srtm.csi.cgiar.org. Access: 05/2009.

Kemp, R., Toms, P., King, M. and Kröhling, D., 2004. The pedosedimentary evolution andchronology of Tortugas, a Late Quaternary type-site of the northern Pampa, Argentina. In: Iriondo, M, Kröhling, D. and Stevaux, J. (Eds.), Advances in the Quaternary of the De la Plata river basin, South America. *Quaternary International*, 114, 101-112.

Kocurek G., Havholm K.G., Deynoux M.and Blakey R.C. 1991, Amalgamated accumulations resulting from climatic and eustatic changes: *Sedimentology* , v. 38, p. 751– 772.

Kröhling, D. 1999. Upper Quaternary of the lower Carcarañá basin, North Pampa, Argentina. Quaternary International. In: Paleoclimates of the Southern Hemisphere. T. Partridge, P. Kershaw and M. Iriondo, (Eds.). *Quaternary International*, 57/58:135-148. Pergamon Press.

Kröhling, D. 2008. Secuencias cuaternarias loess-paleosuelos y arena eólica-paleosuelos de la llanua pampeana: avances en la correlación a escala regional. XVII Congreso Geológico Argentino, Actas 3:1226.

Kröhling, D. and Iriondo, M. 2003. El loess de la Pampa Norte en el Bloque de San Guillermo. *Revista de la Asociación Argentina de Sedimentología*, 10(2):137-150.

Kröhling, D., 2010. Dissipation processes on the Pampean Sand Sea (Late Quaternary) deduced from sedimentological data. 18th International Sedimentological Congress. Abstracts. Mendoza, Argentina.

Kröhling, D., Passeggi, E., Zucol, A., Erra, G., Miquel, S. and Brea, M. 2010 . Multidisciplinary analysis of the last glacial loess at the NE of the Pampean aeolian system. 18th International Sedimentological Congress. Abstracts:518. Mendoza, Argentina.

Krumbein, W. and Sloss, 1955. *Stratigraphy and Sedimentation*. Freeman and Co., San Francisco, 660 pp.

Lancaster, N., 1993. Development of Kelso Dunes, Mojave Desert, California. *National Geogr. Res. Explor.*, 9: 444-459.

Lancaster, N., 1999. Sand seas. In Aeolian Environments, Sediments, and Landforms (A. S.Goudie, I. Livingstone and S. Stokes, Eds.), pp. 49–70. New York, Wiley.

Lancaster, N., 2006. Dune, Aeolian. In: Goudie A.S. (Ed.). *Encyclopedia of Geomorphology. Taylor and Francis e-Library*. London and New York; p: 285-291.

Lancaster, N., 2007. Dune Fields - Low Latitudes. . In *Encyclopedia of Quaternary Science*. Elsevier, p: 626 - 642.

Mc Kee, E. 1979 – A study of global sand seas. Geological Service Professional paper 1052, 429 pp. Washington, USA.

Muller, G. 1967. Methods in Sedimentary Petrology. E. Schweitzerbeistische Verlagsbuchhandlung. Stuttgart, 283 pp.

Nickling, W. 1994. Aeolian sediment transport and deposition. In (K. Pye, ed.) *Sediment transport and depositional processes*. Blackwell Scientific Publications, pp. 293-350. Oxford.

Ollier, C. 1969. *Weathering. Geomorphology* 2. American Elsevier Publ. Co. 304 pp. New York.

Parras, P. 1943. Diario y derrotero de sus viajes. Editorial Solar, Buenos Aires, 251 pp.

Petit-Maire, N., De Beaulieu, J., Iriondo, M., Boulton, G. and Partridge, T. 1999. Maps of the World environments during the two last climatic extremes. Scale 1:10,000,00. Commission ofthe Geological Map of the World/Agence Nationales pour la Gestion des Déchets Radiactifs.,París.

Ramonell, C., Iriondo, M. and Kromer, R. 1992. Guía de Campo No. 1 – Centro-este de San Luis.5°. Reunión de Campo de CADINQUA, 36 pp. San Luis.

Salazar Lea Plaza, 1980. Geomorfología. In *Recursos Naturales de la provincia de La Pampa.* (E. Cano, ed. INTA/UNLP, 319-362. Buenos Aires.

Short, N. and Blair, R. 1986 – Geomorphology from space -A global overview of regionallandforms. NASA SP-486, 717 pp. Washington, USA.

Strahler, A. 1997. Physical Geography. 4th. Edition.John Wiley and Sons, Nes York, 637 pp.

Sun J. and Muhs, D.R., 2007. *Dune Fields - Mid-Latitudes.* In Encyclopedia of Quaternary. Science. Elsevier, p: 607-626.

Talbot, M., 1985. Major bounding surfaces in aeolian sandstones: a climatic model. *Sedimentology* 32:257-266.

Terzaghi, K. and Peck, R. 1963. Mecánica de suelos en la ingeniería práctica. Editorial ElAteneo. 681 pp. Buenos Aires.

Tripaldi, A. and Forman, S., 2007. Geomorphology and chronology of Late Quaternary dune fields of western Argentina. *Palaeogeography, Palaeoclimatology, Palaeoecology*, 251, 300–320.

Tullio, J. 1981. El Cuaternario de la provincia de La Pampa. 14 pp. Ined.

Warren A., 2006. Sand Sea and Dunefield. In: *Encyclopedia of Geomorphology* (A.S. Goudie Ed). Volume 1; 1156 p.

Wentworth, C. 1922. Scale of grade and class terms for clastic sediments. *Geology*, 30: 377-392.

Wilson, I.G., 1973. Ergs, Sedimentary Geology 10, 77–106.

Zárate, M., 2003. The Loess record of Southern South America. *Quaternary Science Reviews* 22, 1987–2006.

In: Sand Dunes
Editor: Jessica A. Murphy, pp. 43-74
ISBN 978-1-61324-108-0

Chapter 2

DESERT SAND DUNES FOR A SUSTAINABLE SULFUR CONCRETE PRODUCTION IN ARID LANDS

Abdel-Mohsen Onsy Mohamed **[*]**
and Maisa Mabrouk El Gamal
Department of Civil and Environmental Engineering,
UAE University, Al Ain, United Arab Emirates

ABSTRACT

The strength and stability of desert sand dunes present challenging problems to geotechnical engineers. To overcome these problems, sulfur concrete as a new material was developed and evaluated in view of its microstructure, physical, thermal, mechanical, hydraulic, and chemical properties. Sulfur is used as the main binding material in sulfur concrete for years. The use of elemental sulfur with aggregates to manufacture sulfur concrete encountered problems due to the complex nature of sulfur chemistry. While initially excellent strength properties were obtained, over a short period of time the material would fail. However, eventually it will convert to a brittle orthorhombic crystal structure with poor strength

[*] Corresponding Author: Prof. Dr. Abdel-Mohsen Onsy Mohamed, Department of Civil and Environmental Engineering, Faculty of Engineering, UAE University, P.O. Box 17555, Al Ain, UAE, Tel: 9713 713 3700, Fax: 9713 762 3154; e-mail: Mohamed.a@uaeu.ac.ae

characteristics. This study has focused on the modification of elemental sulfur; the key to this modification came with the discovery of a novel technique to treat sulfur by preventing the allotropic transformation. The sulfur concrete was prepared from modified sulfur, elemental sulfur, fly ash, and desert sand dunes (soil aggregates). Sulfur concrete specimens were cured in air, water, acidic and saline solutions at different temperatures, ranging from room temperature to 60° C, at different periods. The results indicated that the strength of the manufactured sulfur concrete is about three times higher than that of Portland cement concrete. The results of hydraulic conductivity are ranged between 1.46×10^{-13} and 7.66×10^{-11} m/s making it a good candidate for its potential use as barrier system in arid lands. Durability results indicated that the manufactured material has high resistance to chemical environments being acidic, neutral, and saline and extremely low leached sulfur from the solidified matrix was recorded.

1. Introduction

Sand dunes occur throughout the world, from coastal and lakeshore plains to arid desert regions. In addition to the remarkable structure and patterns of sand dunes, they also provide habitats for a variety of life which is marvelously adapted to this unique environment.

The origin of sand dunes is very complex, but there are three essential prerequisites: (a) an abundant supply of loose sand in a region generally devoid of vegetation; (b) a wind energy source sufficient to move the sand grains; and (c) a topography whereby the sand particles lose their momentum and settle out. Any number of objects, such as shrubs, rocks or fence posts can obstruct the wind force causing sand to pile up in drifts and ultimately large dunes. The direction and velocity of winds, in addition to the local supply of sand, result in a variety of dune shapes and sizes.

The structure and mineral composition of sand grains depends on the geology of the mountains that have been eroded away by wind and water. A typical chemical composition of desert sand dunes obtained from Al Ain city, United Arab Emirates (UAE) is shown in Table 1. It is composed of oxides of silica, magnesium, aluminum and iron in decreasing order. The sand is characterized by a grain size ranging from 0.1 to 1 mm and specific gravity of 2.58. The pore fluid is characterized by a pH of 7.4 and electrical conductivity of 390 μS. Although most dunes are composed of quartz and feldspar grains, the snow-white dunes of White Sands, New Mexico are composed of gypsum and the spectacular black sand beaches of tropical South Pacific islands are made of fine volcanic particles.

Sand dunes can impact humans negatively when they encroach on human habitats. Sand threatens buildings and crops in Africa, the Middle East, and China. Drenching sand dunes with oil stops their migration, but this approach is quite destructive to the dunes' animal habitats and uses a valuable resource. Sand fences might also slow their movement to a crawl, but geologists are still analyzing results for the optimum fence designs. Preventing sand dunes from overwhelming towns, villages, and agricultural areas has become a priority for the United Nations Environment Programmer.

Table 1. Chemical composition of Al Ain, UAE, desert sand dunes

Compound	Weight (%)
SiO_2	67.97
Al_2O_3	2.23
Fe_2O_3	2.35
CaO	10.14
MgO	4.79
SO_3	0.24
K_2O	0.80
Na_2O	0.23
Cl	0.0
Loss on Ignition	11.25

From geo-engineering viewpoint sand dunes are unstable because of the existence of fins sand particles on the surface of the ground. These fine particles are poorly bonded and susceptible to erosion by wind and rain. In addition, the structure of the soil determines its properties such as hydraulic conductivity, porosity, crust formation and load carrying capacity. Weak soils and sand structures cause problems in road and highway construction, steep slopes, water channels, containment of waste, construction, excavation banks, and landing sites for aircrafts, etc.

Extensive research has been carried out to improve soil structure, to reduce soil erosion, to reduce water evaporation, and to increase bonding strength so that stabilized soils could withstand greater loads. Traditional ground improvement/ soil stabilization techniques, which employed by many practicing geotechnical engineers, could not be employed in desert arid regions because of lack of water and high evaporation rates. Therefore, this study shifts its attention to the issue of harvesting desert sand dunes through its utilization as a raw material for manufacturing sulfur concrete that could be utilized in many public works in arid and semi arid lands. Sulfur concrete is a new technology which utilizes sulfur, from oil industry, fly ash, from power generation plants, and desert sand dunes as raw materials. Such new technology will contribute to the increase of sustainability of human activity in relation to building and construction industry.

1.1. Terminology

Mohamed and El Gamal (2010) defined the term "*modified sulfur*" to indicate that chemical additives are added to the elemental sulfur, the term "*sulfur cement*" to indicate the product from the addition of mineral fillers to modified sulfur. Also, they used the term "*sulfur concrete*" when aggregates are added to sulfur cement (Figure 1). Furthermore, to identify the type of chemicals used in modifying the sulfur properties, they attach the name of the used chemicals to the term sulfur concrete, i.e., *bitumen*-based sulfur concrete.

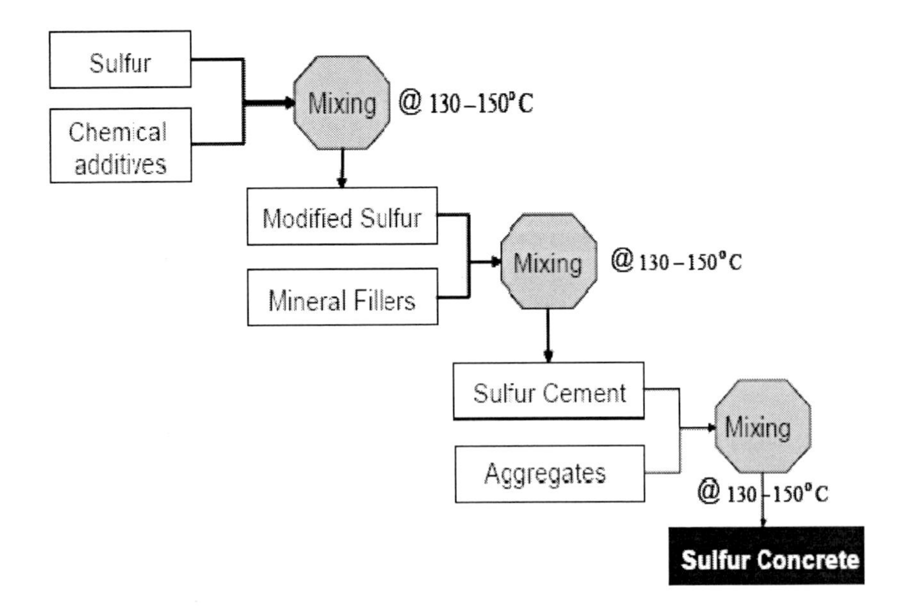

Figure 1. Sulfur terminologies used for sulfur concrete production (Mohamed and El Gamal, 2010).

1.2. Development of Sulfur Concrete

Major advances in the development of sulfur concrete have taken place in the last decades. Research was based on the premise that for sulfur concrete to be a viable construction material, its durability would have to be improved, better sulfur cements would have to be developed, and better mixture designs would have to be formulated for the production of uniform products on a routine basis. The use of the polymer additive provides a simple and effective method of modifying sulfur. Duecker (1934) was the first to modify the sulfur chemically with an olefin poly-sulfur, known commercially as Thiokol that delayed the volume change tendency and the strength loss. The use of sulfur concrete for chemical plants increases to a certain extent, especially for construction in acidic environments in which the durability of the material was excellent. During the following 30 years, much effort was devoted to the study of modifiers and plasticizers for sulfur concrete.

During the 1960s, there was a remarkable investment in environmental protection against discharge of sulfur into the atmosphere, thus making sulfur a surplus commodity on the market, particularly in the United States of America and Canada. This was a crucial point that made the interest in the use sulfur as a structural binder to grow and initiated extensive research programs that became active in the 1970s and focused on various properties of the material, including durability.

In the late 1960s, Dale and Ludwig (1968) pioneered the work on sulfur-aggregate systems, pointing out the need for well-graded aggregates to obtain optimum strength. This work was followed by the investigations of Crow and Bates (1970) on the development of high-strength sulfur-basalt concretes.

At the beginning of the 1970s, successful projects in which sulfur concrete was used as a construction material were carried out on different levels. The United States Department of the Interior's Bureau of Mines and The Sulfur Institute (Washington, D.C.) launched a cooperative Program in 1971 to investigate and develop new uses of sulfur. At the same time, the Canada Center for Mineral and Energy Technology (CANMET) and the National Research Council (NRC) of Canada initiated a research program for the development of sulfur concrete (Malhotra, 1974; Beaudoin and Sereda, 1974).

In 1973 the Sulfur Development Institute of Canada (SUDIC), jointly founded by the Canadian Federal Government, Alberta Provincial Government, and the Canadian sulfur producers, was established to develop new markets for the increasing Canadian sulfur stockpile. In 1978, CANMET and SUDIC sponsored an international conference on sulfur in construction

(Malhotra et al., 1978). Also during this period, a number of investigators (McBee and Sullivan 1979, Vroom 1977 and 1981, Sullivan et al., 1975; McBee and Sullivan 1982; Funke et al., 1982; McBee et al., 1981, 1983; Sullivan, 1986) published number of papers and reports investigated various aspects of sulfur concrete. Most of the studied substances were organic polymers of several types that, upon chemical reaction with sulfur, induced the formation of polysulfide and altered sulfur crystallization (Blight et al., 1978; Jordaan et al., 1978).

McBee et al. (1981; 1982; 1983) published a number of papers and reports dealing with various aspects of sulfur and sulfur concrete. All of these activities led to an increased awareness of the potential use of sulfur as a construction material. As an example, Czarnecki and Gillott (1989) reported sulfur modifiers that improve the ductility and durability of sulfur concrete by strengthening the bonds between the sulfur matrix and the aggregate and between the sulfur crystals in the matrix. Makenya (2001) discussed a proprietary modifier called SRX developed by (Vroom 1992) that cuts the orthorhombic crystalline chains of sulfur into smaller pieces, reducing shrinkage of the cement. Recently, Mohamed and El Gamal (2006; 2007; 2008; 2009) reported a new technology that improved the performance of sulfur concrete products by using bitumen as an additive for sulfur modification.

1.3. Sulfur Concrete Composition

Sulfur concrete is designed to replace Portland cement concrete (ready mix concrete) in many applications. The optimum mixture design for sulfur concrete must take into consideration the properties desired for its specific use. The usual objective of mixed design is to prepare sulfur concrete with the following characteristics: (a) resistance to attack by most acids and/or salt solutions, (b) minimum moisture absorption, (c) mechanical strength properties equivalent to or better than those of Portland cement concrete and (d) sufficient fluidity for good workability.

Because the composition of sulfur concrete is significantly different from Portland cement concrete, the material's selection, mix proportioning, production, and placing are unique. Sulfur concrete row materials are sulfur, chemical additives, mineral fillers and aggregates. These materials are discussed below. Figure 2 shows alternatives for sulfur concrete composition.

The sulfur concrete can be made by adding 5% by weight of a chemical modifying agent to molten sulfur at 130°C. The modified sulfur is then added to aggregate which has been heated to 155°C, and mixed thoroughly and then cast. Usually, the composition of the sulfur concrete is 15%-25% modified sulfur and 75%-85% aggregate by weight (Figure 2). Industrial by products such as fly ash, furnace slag, cement kiln dust, talc, mica, silica, graphite, carbon black, pumice, insoluble salts (e.g., barium carbonate, barium sulfate, calcium carbonate, calcium sulfate, magnesium carbonate, etc.), magnesium oxide, and mixtures have been reported as potential aggregate materials.

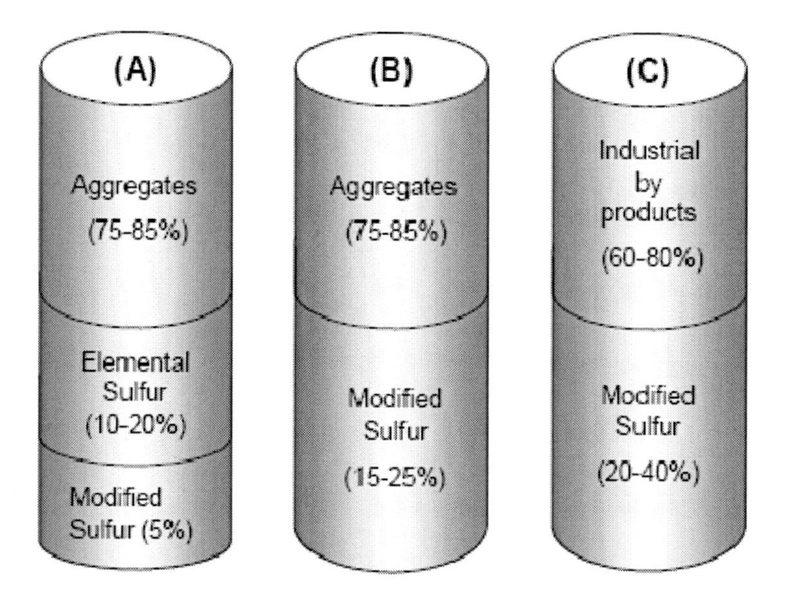

Figure 2. Alternatives for sulfur concrete composition.

Such materials, typically, have a particle size less than 100 mesh (0.149 mm in diameter), as per US Standard Testing Sieves, and preferably, less than 200 mesh (0.074 mm in diameter). In such case, the amount of modified sulfur would increase to 20-40% (Figure 2).

1.3.1. Sulfur

Under standard conditions, elemental sulfur occurs in the environment as soft yellow orthorhombic crystals with a density of 2.07 Mg/m3. The crystalline sulfur has a ring type structure with the formula S8. As the temperature increases, the molten sulfur becomes more viscous. The increase in viscosity is caused by breakage of the ring structure to form chains that may polymerize as the temperature increases to form long molecules or cyclic compounds (Beall and Neff, 2005). The form and purity of the sulfur in sulfur concrete is quite unimportant, providing that the impurities do not represent more than about 4% clay or other water-expansive material. The sulfur may be in either solid or liquid (molten) form.

1.3.2. Chemical Additives

Sulfur modifiers are required to keep the sulfur in a preferential crystalline form that resists concrete fracturing, through the allotropic transformation during cooling, resulting in sulfur concrete that is not highly stressed and is much more durable. For example, Czarnecki and Gillott (1989) described sulfur modifiers that improve the ductility and durability of sulfur concrete by strengthening the bonds between the sulfur matrix and the aggregate and between the sulfur crystals in the matrix. The various types and concentrations of sulfur modifier in sulfur concrete have different degrees of resistance to environmental conditions.

1.3.3. Mineral Fillers

Mineral filler with aggregates and binder are necessary for the assurance of maximum density and limited absorption (Weber el al, 1990). The main functions of mineral fillers are to:

a) Control the viscosity of the fluid sulfur-filler paste, workability, and bleeding of the hot plastic concrete.
b) Act as a thickening agent that reduces the separation tendency and helps to produce homogeneous products.
c) Provide nucleation sites for crystal formation and growth in the paste and to minimize the growth of large needle-like crystals.
d) Fill the voids in the mineral aggregate that would otherwise be filled with sulfur and to reduce shrinkage during hardening.
e) Act as a reinforcing agent in the matrix and, hence, increase the strength of the formation.

Therefore, to meet these functions, the filler must be reasonably dense-graded and possibly finely divided to provide a large number of particles per unit weight, especially to meet the second function, provision of nucleation sites.

Materials such as crushed dust, silicate flour and other inorganic fillers, such as fly ash, have been used as mineral fillers. The amount of filler required in the mix, although typically 5% percent by weight will ultimately be determined by the presence of < 200 mesh (75μm) sized fines in the aggregate source. As with the aggregate, the mineral filler should be tested for chemical resistance to confirm its suitability for the intended environment (Crick and Whitmore, 1998, Mohamed and El Gamal, 2009).

1.3.4. Aggregates

Aggregate is crushed mineral material, usually natural rock of different types. The aggregates used in Portland cement concrete are similar to those used to manufacture sulfur concrete. Three size fractions of aggregate are generally used in preparing the dense-graded product: coarse aggregate, fine aggregate and mineral filler. Waste solids, such as fly ash, steel slag, and cement kiln dust can be used as aggregate or filler in sulfur concrete. The quality of the aggregates has a substantial effect on the properties and durability of the finished sulfur concrete.

The selection of quality aggregates, which will be appropriate for each particular application, is necessary for a sulfur concrete material. To meet the requirement of durability, cleanliness and limits of harmful substances, the composite aggregates must meet the ASTM C33 specification. As per the ACI-guide for mixing and placing sulfur concrete (1993), aggregates should conform to and be classified under the following requirements:

a) *Aggregate gradation*: Dense graded aggregates are used in the production of sulfur concrete to minimize binder requirement.
b) *Corrosion resistant aggregates*: Should not show any effervescence when tested in acid of a given concentration and at a given temperature. Aggregates that are insoluble in acids such as quartz should be used for preparing acid-resistant sulfur concrete; salt-tolerant sulfur concrete can be prepared from crushed quartz or limestone. Crushed rock should be used because it bonds better with sulfur than weathered, rounded sand and gravel.
c) *Moisture absorption of aggregates*: Should be highly impervious and highly resistant to freeze-thaw. Clay contained within the solidified sulfur concrete is believed to have an absorptive capacity, which will allow water to permeate through the material. When clay absorbs water, expansion occurs and will result in deterioration of the product. Therefore, clay containing aggregates should not be used in producing sulfur concrete. Hence, porous aggregates, in general, should not be used.

2. Bitumen Modified Sulfur Concrete (BMSC)

2.1. Production of BMSC

Bitumen modified sulfur concrete (BMSC) consisting of elemental sulfur, modified sulfur, fly ash and desert sand was prepared according to the procedure described in Mohamed and El Gamal (2007). The physical additives or aggregates (fly ash and desert sand) were heated in an oven to 170-200°C for a period of two hours. The specified amount of sulfur was melted in a heated mixing bowl that was placed in oil bath with controlled temperature in the range of 132-141°C. Fly ash was then transferred to the heated mixing bowl and properly mixed with the molten sulfur for about 20 minutes to insure a complete reaction between sulfur and fly ash. Modified sulfur was added and mixing continued for additional 5 minutes. Finally, desert sand was added and mixing was continued for about 20 minutes more. In this process, the proportions of 0.25 wt% for modified sulfur, 0.9 for sulfur to fly ash, one (1) for sulfur to sand were used. At this stage of preparation, the mixture is viscous and can be easily bored into specified moulds for casting. For specimen preparations, cubical steel molds were used. The molds were preheated to approximately 120°C before adding the viscous BMSC. During the boring of BMSC into the specified molds, the mixture was compacted for 10 seconds on a vibrating table. The surface of each specimen was then finished and the molds were placed in an oven at 100°C with a controlled cooling rate of 5° per minute. After 24 hours, specimens were de-molded.

2.2. Effect of Sulfur Ratio and Loading

The desirable balance between sulfur binders composed of elemental sulfur and modified sulfur cement, with the ratio of aggregates have been studied. Sufficient workability of the mixture must be ensured to allow homogenous mixing and operation of equipment within mechanical limit. Basic mechanical properties were measured for the sulfur concrete samples to ensure that the developed materials met the properties found in literature for these materials. Figure 3 shows the desired optimum amount of sulfur to fly ash ratio according to the mechanical strength of the mortar. The obtained strength tended to increase as the sulfur/fly ash ratio increases up to 0.9, where all particles are coated by a thin layer of sulfur. However, with large sulfur addition, the compressive strength decreased, because of further increment of sulfur content increases the thickness of sulfur layers around the aggregate particles that leads to the increase of the brittleness of the formed composite material (Mohamed 2002, 2003, Mohamed et al. 2002, 2003).

To have proper criteria for evaluating the difference between the performance of elemental sulfur concrete and BMSC, samples were prepared using various sulfur loadings. The effect of the modified sulfur incorporated into the mortars is shown in Figure 4. The compressive strength decreased linearly with increasing amounts of modified sulfur due to the partial inhibition of the crystallization through addition of modified sulfur. These results are in agreement with those reported by Vroom (1981). Mortars with modified sulfur show higher viscosity than unmodified one. This fact has an important effect on crystallization of sulfur. In a more viscous liquid, the growth of the crystals will be more difficult and slow, causing partial reduction in compressive strength.

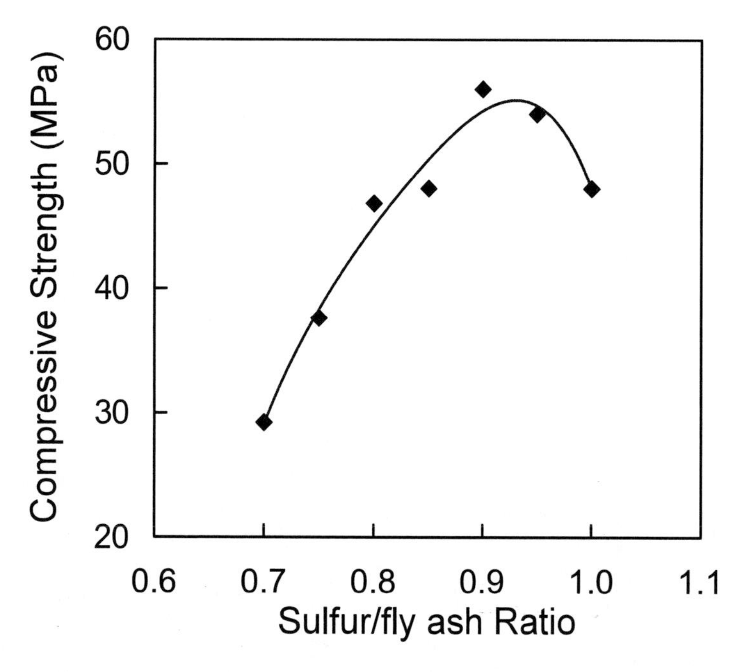

Figure 3. Effect of sulfur ratio on the compressive strength of sulfur concrete.

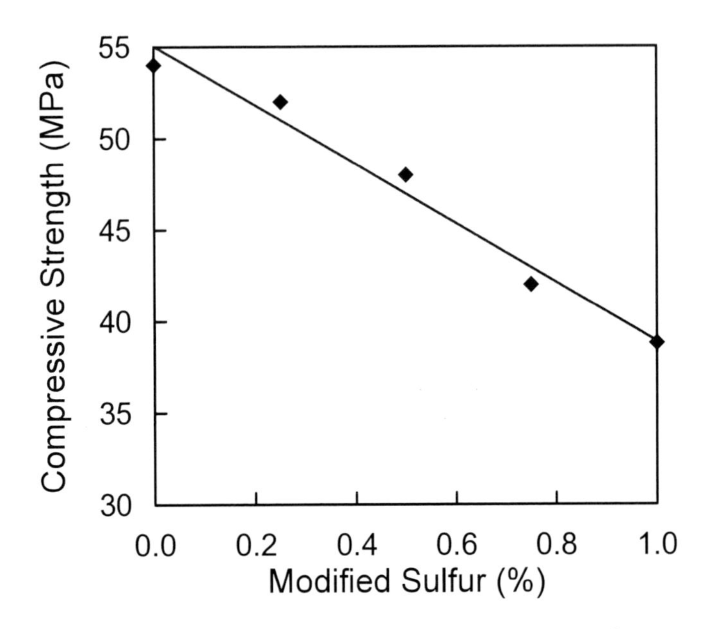

Figure 4. Effect of sulfur loadings on compressive strength.

2.3. Microstructure Characterization

The microstructure characterization, performed by scanning electron microscopy (SEM), provides important data on sulfur concrete studies. SEM Images (Figures 5 (a) and (b)) show sulfur binding the aggregates with filling the inner spaces in such a way that the voids are discrete and homogeneous. This in turn increases the strength of the developed matrix.

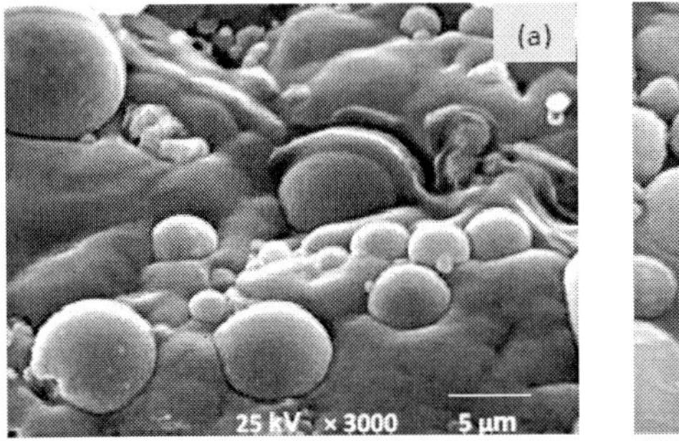

Figure 5. SEM micrographs of BMSC show (a) sulfur aggregates binding system, and (b) particle shapes, sizes and voids.

2.4. Strength Development

The compressive strength of cylindrical sulfur concrete specimens was performed according to ASTM C39 (1986). BMSC samples were prepared at 135°C, cured in the oven with gradual cooling rate of 5 degrees/min, de-molded after 24 hours, and then cured at the tested temperature. The raw materials used in these preparations possess some specific known characteristics. The main aspects of the concrete performance that will be improved by the use of fly ash are increased long-term strength and reduced hydraulic conductivity of the concrete resulting in potentially better durability. The use of fly ash in concrete can also overcome some specific durability issues such as sulfate attack and alkali silica reaction. The desert sand was found to be of good quality showing a low concentration of impurities and grains with an irregular geometry.

This quality of sand grains may reduce the workability of the mortar but on the other hand enables the molten sulfur to adhere more easily on the surface of the sand grains (Gemelli et al., 2004). The molten sulfur acts as binder for these aggregates. Mortar with modified sulfur shows high viscosity, which has an important effect on the crystallization of sulfur. The modified sulfur is very efficient in binding and strengthening the aggregates and has been found to impart an extra durability to the final BMSC.

Compressive strength development with time, for BMSC specimens cured in an oven with gradual cooling rate of 2 degrees/min, de-molded after 24 hours, and stored in incubators at 40°C, is shown in Figure 6. It is clear that BMSC has developed about 76% of its ultimate compressive strength within one day and 97% after 3 days.

However, at later times up to 42 days, there was no clear trend on the compressive strength development suggesting that the maximum strength was developed during the early days. Similar results were reported by many researchers. For example, Vroom (1981) reported that 80% of the ultimate concrete strength was developed in one day, and virtually 100% of the ultimate strength was realized after four days. McBee et al. (1983) showed that the sulfur concrete developed about 70% of its ultimate strength within a few hours after cooling, 75 to 85% after 24 hours at 20°C, and the ultimate strength was commonly obtained after 180 days at 20°C.

To evaluate the effect of curing temperatures, BMSC specimens were prepared and cured for known temperatures and time periods in specified environment (air and de-ionized water). Specimens were cured at 24, 40 and

60°C. Figure 7 shows the variations of compressive strength in the air and de-ionized water with time at different temperatures.

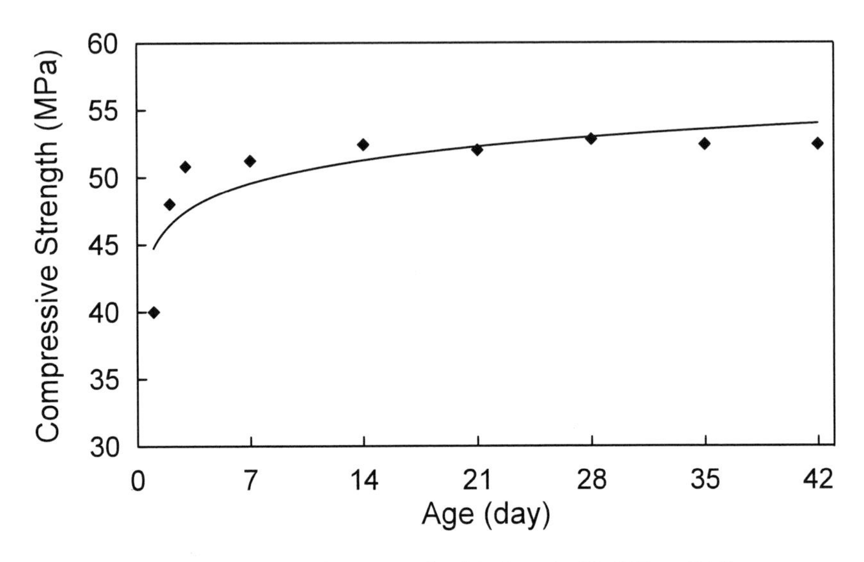

Figure 6. Variations of compressive strength with time for BMSC at 40°C.

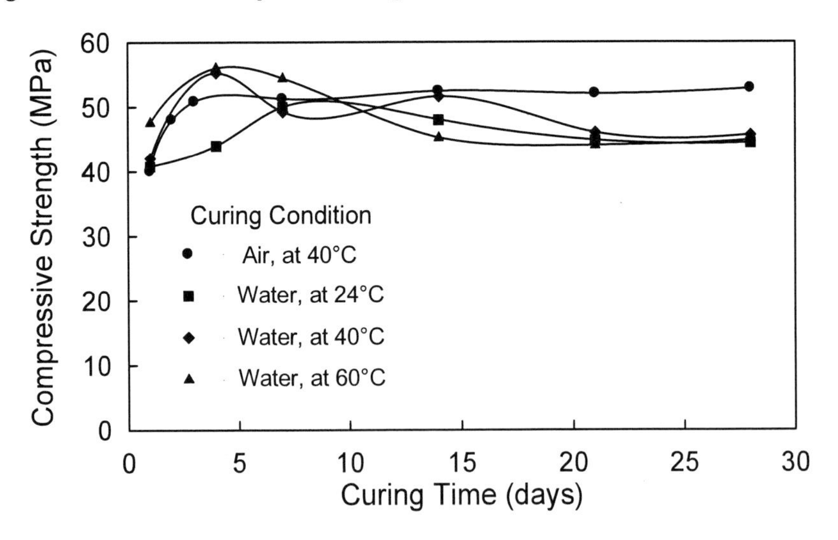

Figure 7. Influence of curing condition on the strength of BMSC samples.

The results, shown in Figure 7, indicate that, for the first 3 days of immersion, the higher the water temperature and strength gain. It is also shown that after 3days the strength loss occurred up to 15 days and then the strength remains nearly constant up to 28 days. A comparison between the losses in strength, after 28 days of BMSC cured in air and BMSC immersed in water, indicates that curing of BMSC in water resulted in some decrease in compressive strength.

It is clear that BMSC reaches hardening and gains its strength over a short time, resulting in high strength material with an average compressive strength of 54 MPa. The strength increase could be attributed to the formation of cementing agents due to water availability.

Strength reduction was investigated via microstructure analysis of the tested specimens. Figures 8(a) and (b) show SEM micrograph of fractured surface of BMSC and its x-ray mapping for the sulfur elements of the same area for specimens cured in air and de-ionized water after aging time of 28 days at 40°C.

The fractured surfaces of BMSC reveal a different morphology. BMSC cured in the air dry condition results in a homogenous sulfur distribution with good binding and coating of the aggregates as shown in Figure 8(a). For specimens cured in de-ionized water, one observes that no homogenous distribution of sulfur, where parts of the aggregate surfaces were uncoated with sulfur as shown in Figure 8(b).

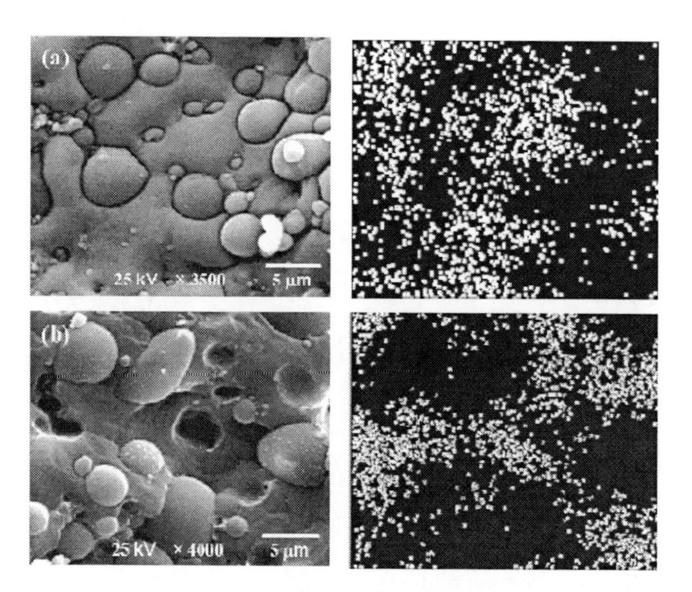

Figure 8. SEM micrograph of fractured surface of BMSC and its x-ray mapping for the sulfur elements of the same area; (a) air cured (b) water cured, after age time of 28 days at 40°C.

This observation appears to explain the small strength reduction of BMSC cured in water, which is in complete agreement with reported studies by various investigators.

2.5. Thermal Stability

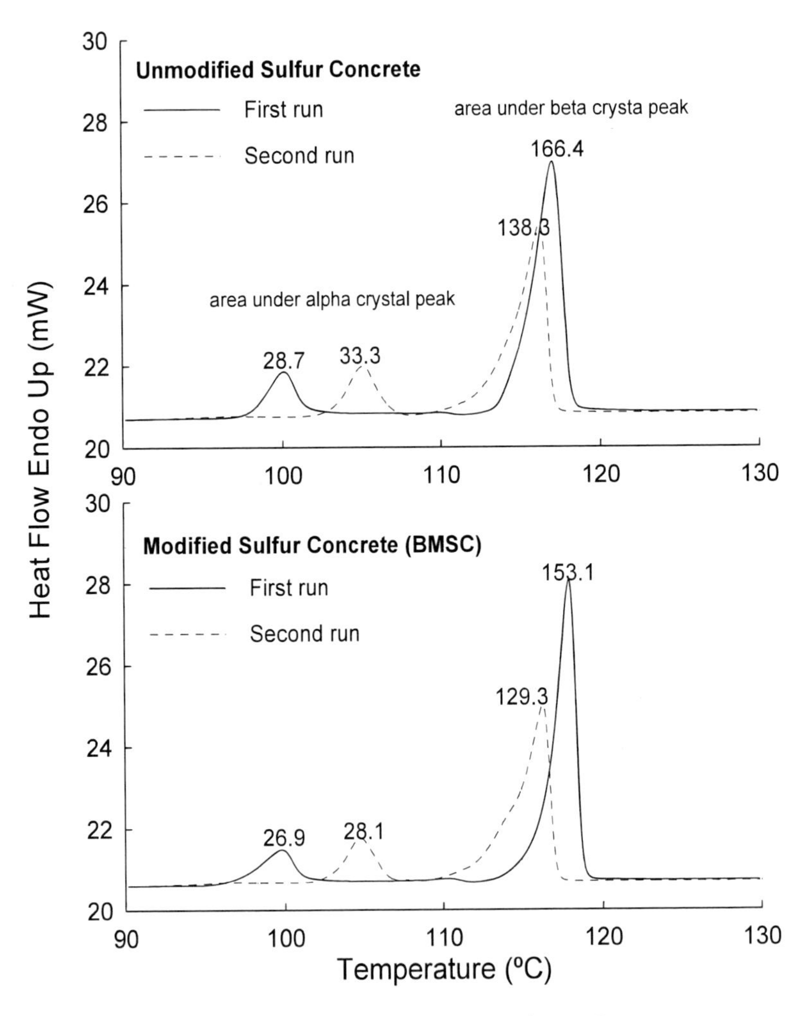

Figure 9. DSC curves for unmodified sulfur concrete and BMSC.

A differential scanning calorimeter (DSC) was used for measurements of heat capacity, through phase transitions on heating. Ten (10) mg of the tested sample was heated up to 150°C, with heating rate of 5°C/min. The sample was allowed to be self cooled to room temperature for 24 hours, and then the sample was reheated up to 150°C. Sulfur crystal forms are obtained using the DSC as shown in Figure 9 for elemental sulfur concrete and bitumen modified sulfur concrete (BMSC). The addition of modified sulfur to the mix at small percentages not only directly affects the crystal form of sulfur (making it crystallize in a structure different than the stable one), but also alters the behavior in the hot liquid mix period.

Crystallization of sulfur in this case is controlled by the relative percentages and distribution of space between aggregates and sulfur binder. The effect of modification seems to be related to the increase of degree of sulfur polymerization. Sulfur itself possesses a tendency to polymerize; the modified sulfur would thus increase this tendency or maintain it for a longer time. In a more viscous liquid, and when the molecules are more polymerized, the growth of the crystal will be more difficult. This fact has an important effect in the crystallization of sulfur. The DSC results for BMSC show significant reduction in the (α) and (ß) sulfur crystal forms, which are illustrated by the reduction of the areas under melted peaks, confirming that BMSC has a low order of crystalline structure compared to elemental sulfur concrete.

To evaluate the effect of both temperature and curing time, compressive strength was measured for BMSC samples cured at different temperatures of 26, 40, 60, 80 and 100°C and for different periods of 1, 2, and 7 days. The variations of compressive strength with temperature and curing time for BMSC are shown in Figure 10. The results indicate that for the same curing time as temperature increases, compressive strength decreases. A systematic reduction in compressive strength with temperature was obtained specially at high temperatures (80 and 100°C). For the same temperature, as curing time increases, compressive strength increases. At temperatures below 60°C, the strength increase with time ranges from 20 to 30% while at higher temperatures, the rate of strength increase with time is low. Similar results were reported by McBee and Sullivan (1979) who found that the strength gain is slower at elevated temperatures and faster at lower temperatures. The results also indicate that the compressive strength of BMSC for temperatures below 60°C is 20 to 30% more than Portland cement concrete and for higher temperatures (60 to 100°C), the compressive strength remains within the same range.

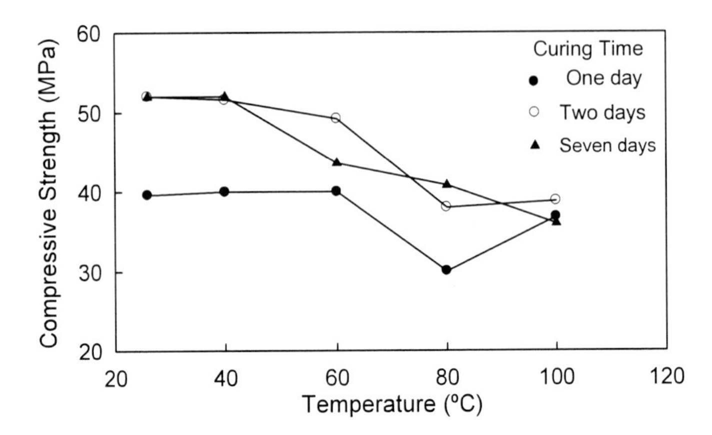

Figure 10. Variations of compressive strength with temperature and curing time for bitumen modified sulfur concrete (BMSC).

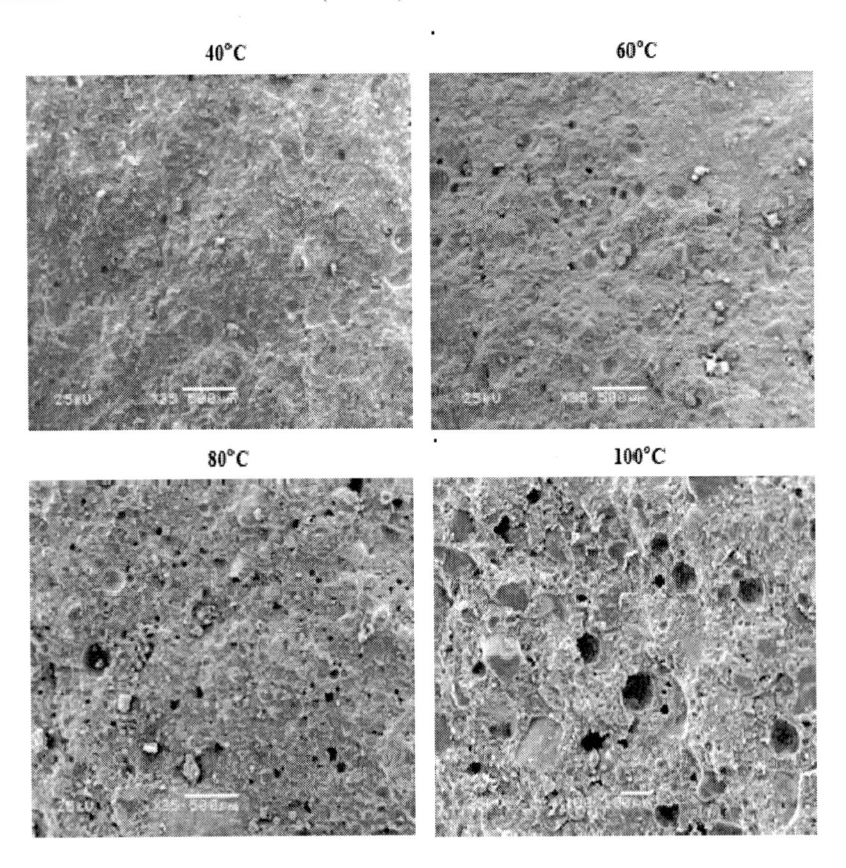

Figure 11. SEM images for bitumen modified sulfur concrete at different temperatures.

Because mechanical strength is directly related to the defects in mortar microstructure it is important to characterize such defects using SEM. Microscopic studies were determined from a section cut at a distance of 5 mm from the surface of BMSC samples.

Figure 11 shows a comparison between the mortar microstructure of BMSC cured after seven days at different temperatures of 40, 60, 80 and 100oC. It is obvious that for BMSC cured at 40 and 60°C, the voids entrained within the BMSC are small and discontinuous. However, for BMSC cured at 80 and 100°C, the voids are large in sizes without any noticeable cracking. The voids in BMSC serve as sites for stress relief and for improving the durability of the material (McBee et al., 1983). Also, the presence of voids in BMSC reduces the quantity of sulfur cement required to coat the mineral aggregate thereby minimizing the cement related shrinkage.

2.6. Durability

Durability of BMSC in different chemical environments and temperatures has been studied. Specimens were immersed in: (a) distilled water at different temperatures of 24°, 40°, and 60°C; (b) 3% saline solution at 40° and 60°C; and (c) 70% sulfuric acid solution at 40°C. All experiments were run for 28 days.

2.6.1. Short - Term Hydro-Mechanical Behavior

(i) Moisture Absorption

The water absorption percent for PCC in water and saline solution reached 12.5 and 7.7%, respectively, which are considered high. However, for BMSC, the recorded water absorption values for water and saline solutions were 0.17 and 0.25%, respectively, indicating that BMSC has low water absorption characteristics in comparison with PCC. To explain such results, one has to realize that the existing sulfur and the added bitumen repel water penetration because they are hydrophobic in nature. Also, the majority of the matrix is composed of sulfur coated materials (fly ash and desert sand) and sulfur in the voids between particles. This in turn will lead to what is known as BMSC impermeability characteristics (Darnell, 1991). These results are in agreement with recommended specifications (ACI Committee 548, 1993), in which the maximum moisture absorption of BMSC should be less than 1% for coarse aggregates and less than 2% for fine aggregates.

(ii) Compressive Strength Development in Acidic Environment

The potential deterioration of BMSC specimens in acidic solutions was evaluated via compression test results after specimen's immersion in different concentrations of sulfuric acid solutions for one week. The results shown in Figure 12 indicate a slight decrease in compressive strength with increase of acid concentration. This is ascribed to the consumption of small amounts of sulfur due to the slow reaction between sulfur and sulfuric acid. Hence, the sulfur binding strength between aggregates is reduced leading to the decrease in BMSC strength. BMSC maintains its high strength, even in a concentrated acidic environment (98% H2SO4). This is an indication that BMSC is acid resistant.

The behavior of both BMSC and PCC in acidic environments was evaluated. Figure 13(a) shows the loss in weight for concrete samples after immersion in different sulfuric acid concentrations (20, 40, 70 and 98% H2SO4), for 1 and 7 days. There is no significant loss in weight for BMSC samples especially that immersed in dilute sulfuric acid, and a small loss in weight (0.19%) was detected after immersing the samples for 24 hours in 98% H2SO4 solution. No more loss in weight was observed for immersed BMSC samples for 7 days at different sulfuric acid concentration.

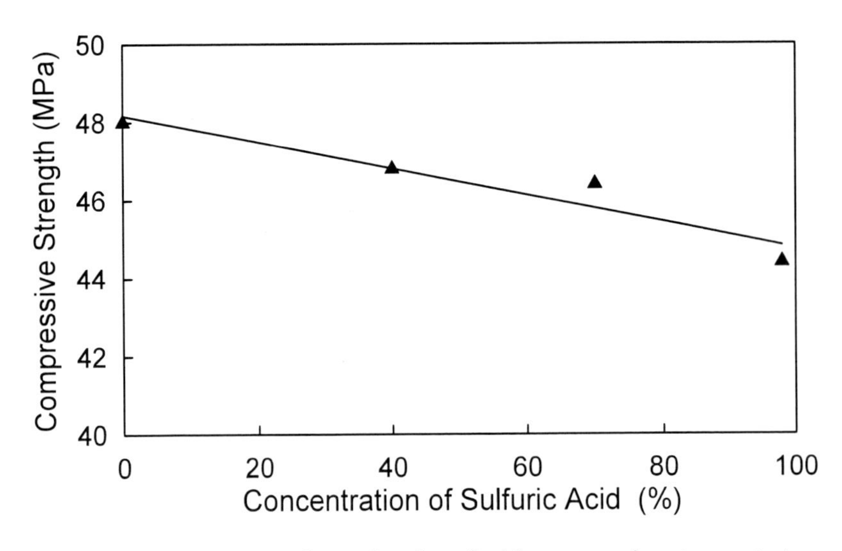

Figure 12. Compressive strength as a function of acid concentration at a constant temperature of 24°C and curing time of 7 days.

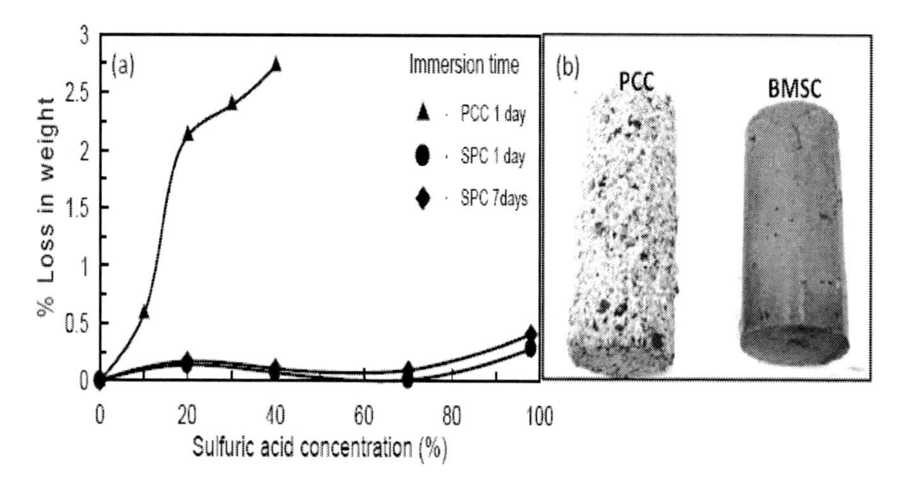

Figure 13. (a) Percent loss in weight as a function of sulfuric acid concentration, and (b) surface morphology of PCC and BMSC due to sulfuric acid attack.

For PCC, the behavior with respect to acid attack is different. It is known that PCC is highly alkaline material and is not very resistant to attack by acids. Immersed PCC specimens in sulfuric acid solutions showed high effervescence, and a loss in weight of 2.7% for samples immersed 24 hours in 40% H2SO4 solution was recorded. In addition to the observed weight changes, the color and surface morphology were observed. Figure 13(b) shows the corroded PCC surface after 24 hours of immersion in 40% H2SO4 solution, which becomes soft and white due to formation of calcium sulfate hydrate. This indicates that BMSC offers better protection to acid attack compared to PCC.

(iii) Compressive Strength Development in Saline Environment

The BMSC durability in saline environments, after immersion of BMSC specimens in NaCl solution for 7 days at a constant temperature of 24°C, was evaluated. The results indicated that there is no loss in weight after immersion. Higher resistance to saline environment was achieved even with a small loss in compressive strength at high salinity as shown in Figure 14.

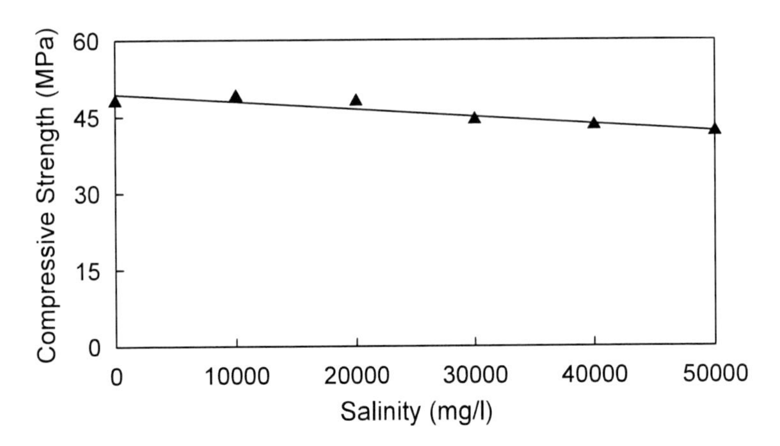

Figure 14. Compressive strength for BMSC as a function of sodium chloride concentration, after immersion of 7 days at a constant temperature of 24°C.

Concerning the microstructure of BMSC specimens, Figure 15 shows a high-magnification of surface image, for the contact regions between sulfur and physical aggregates (fly ash and desert sand). The growth of sodium chloride crystals leads to a partial detaching between sulfur and physical aggregates that, in turn, results in compressive strength reduction. The chemical analysis performed by ICP, for BMSC specimens immersed for 7 days in 5% NaCl solution, indicates the presence of 0.049% Na atoms at the surface (i.e., 0.125% NaCl molecules).

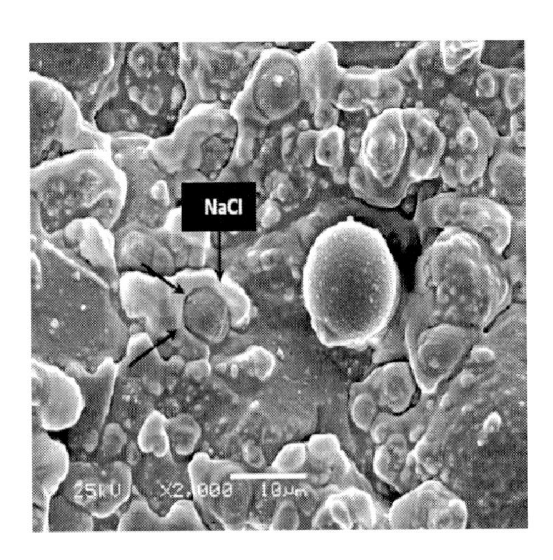

Figure 15. SEM micrograph of BMSC, after immersed of 7 days, in 5% sodium chloride solution, at 24°C.

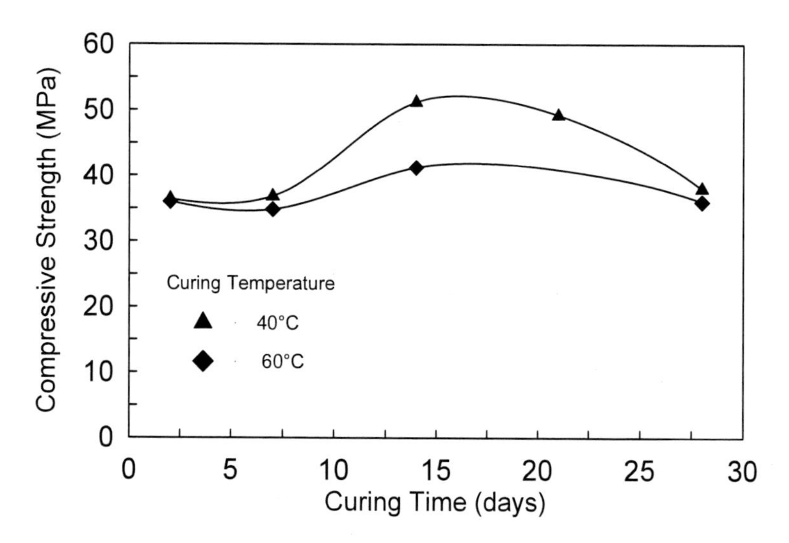

Figure 16. Change in compressive strength of BMSC, cured in 3% sodium chloride solution at 40 and 60°C, with curing time.

However, for BMSC section at depth of 10 mm from the surface, the analysis does not show any sodium atoms indicating that NaCl penetration is only limited to the outer surface of BMSC specimens. The implication of such finding is that BMSC is a corrosion-resistant material and could be reinforced. In addition, hydraulic conductivity measurements, for BMSC specimens immersed in 3 % NaCl solution, indicated that BMSC has low values in the range of 10-11 to 10-13 m/s that are less that for the case of de-ionized water by about two orders of magnitude. Such low hydraulic conductivity makes BMSC a good candidate for stabilization of hazardous waste and barrier system design (Mohamed and El Gamal, 2008, 2009). The variations of compressive strength with time for BMSC samples cured in 3% NaCl solution at 40 and 60°C are shown in Figure 16. It is clear that the higher the water temperature the lower the strength gain is. Therefore, BMSC specimens have maintained their acceptable strength, indicating good resistance to saline environment.

(iv) Hydraulic Conductivity

Measured hydraulic conductivities of BMSC and PCC indicated that BMSC has low water permeation. Under a pressure of 2.2 MPa, BMSC has a hydraulic conductivity in the order of 1.456×10-13m/s, whereas PCC has a hydraulic conductivity of approximately 8.39 ×10-8 m/s after being immersed in water. It is interesting to note that some researchers (Vroom, 1998) have evaluated the volume of pore spaces of BMSC and PCC and have indicated that they have approximately the same volume; however, the pores in BMSC concrete are not connected providing low hydraulic conductivity characteristics, whereas the pores of PCC concrete are connected. It should be noted that the hydraulic conductivity of a material is highly dependent on the size of pore spaces, degree of connectivity between pores, grain shape, degree of packing, and cementation (Yong et al 1992). To evaluate the effect of acid attack on the hydraulic conductivity, BMSC specimens were immersed in 98% sulfuric acid, 50% phosphoric acid, 30% boric acid and 10% acetic acid, at 24°C. Experimental results revealed that hydraulic conductivity values of BMSC specimens are in the range of 10-11-10-13 m/s, indicating that BMSC is an impermeable material. Table 2 summarizes the experimental results of the hydraulic conductivity, loss in weights due to chemical reaction, and compressive strength loss. These data have revealed that BMSC exhibits high resistance to aggressive acidic environment. PCC under the same conditions, in most cases, were destroyed. Evaluation of the effect of the acid concentration on the durability of BMSC was studied through immersion of BMSC specimens in 20, 40, 70, and 98 wt% sulfuric acid solutions, for 7 days at 24°C. Specimens were then washed, dried and examined by SEM as well as chemical analysis using energy dispersive x-ray spectroscopy (EDX).

Table 2. Effect of acid type on hydraulic conductivity measurements, weight loss and compressive strength loss, of BMSC after 7 days immersion in corrosive acids (Mohamed and El Gamal, 2009)

Acid type	Hydraulic conductivity (m/s)	Weight Loss (%)	Strength Loss (%)
water	1.456×10^{-13}	0.00	0.0
98% Sulfuric	7.660×10^{-11}	0.23	13.5
50% Phosphoric	3.103×10^{-12}	0.08	7.9
30% Boric	8.176×10^{-13}	0.07	4.0
10% Acetic	2.196×10^{-12}	0.14	16.0

The results shown in Table 3 indicated that BMSC samples were composed mainly of silicon and sulfur containing compounds and various metals such as calcium, aluminum, iron, and potassium.

Table 3. Effect of acid concentration on the % element of BMSC composition, the data obtained from the EDX (Mohamed and El Gamal, 2009)

Element%	Sulfuric Acid Concentration			
	20%	40%	70%	98%
O	34.9	26.6	20.44	25.06
Al	7.64	11.98	8.06	4.84
Si	28.21	29.76	21.93	11.51
S	24.6	26.85	44.49	56.01
K	0.93	0.89	1.25	0.54
Ca	1.87	1.02	0.39	0.42
Fe	1.84	2.89	3.43	1.63

The EDX observation revealed that with an increase of sulfuric acid concentration, the percentage of elemental sulfur increases because of the formation of metal sulfates due to the reaction of the basic oxides (in the fly ash and desert sand) with sulfuric acid. It is known that most bases dissolve in water releasing hydroxide ions (OH^-) that react with acids to form salts. A calcium oxide base accepts hydrogen ions; therefore, one could say that a base is a proton acceptor. With an increase of sulfuric acid concentration, a reduction in the capillary porosity of the system could be obtained due to salt formation that precipitated on the BMSC surface and in the pore spaces.

2.6.2. Long-Term Hydro-Mechanical Behavior

BMSC specimens were further tested to determine their durability in hydrates and saline environments after one year, at room temperature (24±2°C). Figure 17 shows the compressive strength variations after one year of immersion of specimens in distilled water and in different saline solutions. The results indicated that there is no loss in weight and no adverse effects in compressive strength. BMSC is corrosion-resistant and could be used in hydrated and salt environments. It did not exhibit any deterioration, only limited compressive strength loss was observed.

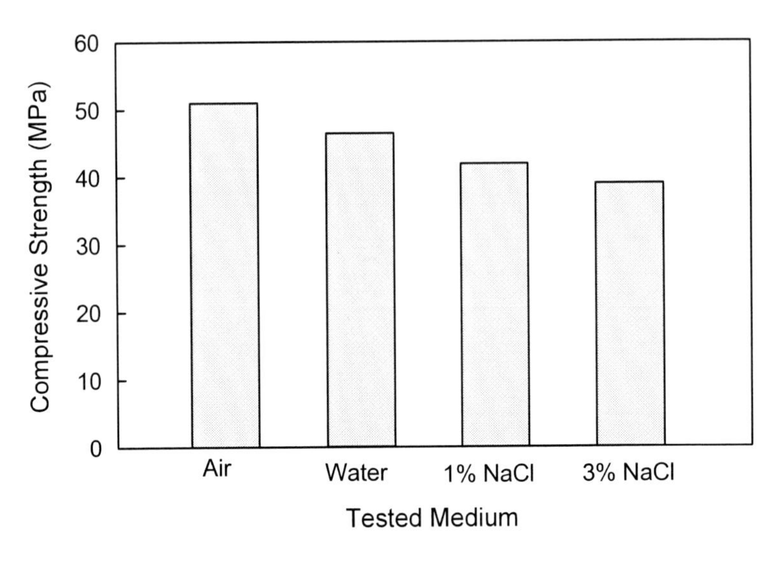

Figure 17. One year immersion test of BMSC in different saline solutions, at 24°C.

2.7. Potential Ecological Effects of BMSC

Sulfur-based concrete resembles PCC in general composition except that a modified sulfur binder replaces the PC. PCC is considered toxicologically nearly inert and usually does not cause more than physical alteration to habitats where it is used. Chemical investigations on possible leaching of sulfur from BMSC specimens immersed in de-ionized water were conducted; there were no significant signs of strength loss or deterioration under actual operating conditions. Decomposition products were measured as sulfates, using Inductively Coupled Plasma–Atomic Emission Spectrometry (ICP-AES). Figure 18 shows the percentage of leached sulfur during 1-year. The results indicated that at room temperature, BMSC was very stable and insoluble in water through the test duration, where no pH change or weight loss was observed. As the duration time of leached test increases, it was found that; the total leached sulfur in water was nearly steady. These results indicate that there was little or no loss of sulfur from the structures. Elemental sulfur has a low aqueous solubility and sulfur concrete has a low water content and hydraulic conductivity, reducing the possibility of significant leaching of sulfur (Mohamed and El Gamal, 2009; Wrzesinski et al., 1988). The calcium salts in PC can leach from the concrete and increase the pH of the surrounding. If sulfur leaches from sulfur concrete, it will have the opposite effect, compared to PCC, on the soil pH in the immediate vicinity of the concrete structure.

Elemental sulfur (the form of sulfur in sulfur cement and concrete) is oxidized in soil to sulfate, which reacts with soil water to form sulfuric acid, reducing soil pH. Warren and Dudas (1992) reported pH values below 3.0 in soils adjacent to an elemental sulfur stockpile. Because soil pH near sulfur concrete structures will not be as low as the pH values measured near the sulfur stockpile, it is doubtful that there will be substantial mobilization and redistribution of metals in the soil (Beall and Neff, 2005).

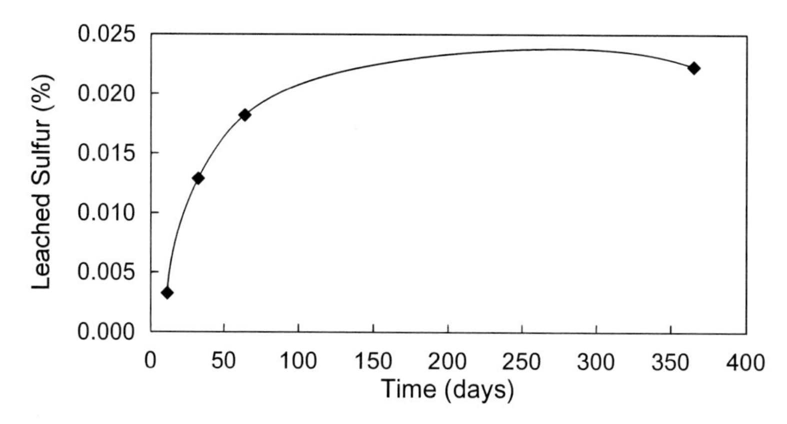

Figure 18. Variations of the amount of sulfur leached from BMSC in de-ionized water at 24°C, as function of time.

Kalb et al. (1991) showed by TCLP analysis that leaching of metals from incinerator fly ash was reduced to acceptable levels by encapsulating the fly ash in modified sulfur cement. Thus, sulfur concrete is unlikely to leach metals even under severe environmental conditions (Beall and Neff, 2005). Modified sulfur binder in most sulfur cement and concretes are considered to be insoluble in water and are unlikely to leach from the concrete.

CONCLUSION

This study was aimed to develop a method for utilization of desert sand dunes as a raw material for manufacturing sulfur concrete having satisfactory physical, thermal, mechanical and environmental properties. Sulfur concrete test results revealed that, the products are highly durable and could be used as construction materials in locations within industrial plants or other locations where acid and salt environments result in premature deterioration and failure of Portland cement concrete. Scan micrographs indicated that the produced matrix is very dense with clear penetration of modified sulfur between the aggregates. 80% of the maximum strength was achieved in a little than one day and the product is ready for use in less than one day. This fast curing contributes to the shortening period of construction. Since hydraulic conductivity is an important parameter in some geotechnical applications, the developed matrix was subjected to aggressive chemical environments and its hydraulic conductivity was evaluated. The results indicated that hydraulic conductivities in the order of 1.0E-11 to 1.0E-13 m/s were achieved indicating that the manufactured sulfur concrete is impervious to water and corrosive environmental solutions. From the thermo-mechanical and hydro-chemical behavior viewpoints, the results clearly indicated that the developed material is viable and could be used for containment of hazardous waste in arid lands because of its (1) fast hardening, in less than a day; (2) high strength, two to three times of Portland cement concrete; (3) high resistance to chemical environments being acidic, neutral, and alkaline; (4) very low leachability of sulfur from the solidified matrix.

Sulfur concrete has the following useful characteristics: (a) low hydraulic conductivity and porosity, (b) mechanical properties, such as tensile, compressive, and flexural strengths, and fatigue life are greater than those obtained with normal Portland cement concrete, (c) comparable density with that of Portland cement concrete, (d) excellent resistance to attack by most acids and salts, some at very high concentrations, also resists corrosive electrolyte attack, (e) can be placed year-round, (f) compositional materials could be recycled and used again, (g) impurity of used materials does not have any effect on the final strength properties, (h) no need of water for production, (i) possibility of color coding without any problem, (j) can be utilized as construction materials for floors, walls and sump pits in the chemical, metallurgical, battery, fertilizer, food, pulp and paper industries, and (k) can be utilized as road paving and bridge decking where salt corrosion problems are encountered.

REFERENCES

ACI Committee 548, (1993) "*Guide for mixing and placing sulfur concrete in construction (ACI 548.2R-93),*" American Concrete Institute, Farmington Hills, Mich.

Beall, P. W. and Neff, J. M. (2005) "Potential non-traditional uses of by-product EandP produced sulfur in Kazakhstan." SPE 94177 This paper was prepared for presentation at the 200 5 SPE/EPA/DOE Exploration and Production Environmental Conference held in Galveston, Texas, U.S.A., 7 –9 March 2005.

Beaudoin, J.J., and Sereda, P.J. (1974) "*The freeze-thaw durability of sulfur concrete,*" Building Research Note, Division of Building Research, National Research Council, Ottawa, 53.

Blight, L.; Currell, B.R.; Nash, B.J.; Scott, R.A.M., and Stillo, C. (1978) "Preparation and properties of modified sulfur systems," In *New Uses of Sulfur -II, Advances in Chemistry Series* No. 165 (American Chemical Society), pp.13-30.

Crick, S. M., and Whitmore, D. W., (1998) "Using sulfur concrete on a commercial scale," *Concrete International*, Vol. 20, February, 1998, pp. 83-86.

Crow, L.J. and Bates, R. C. (1970) "*Strength of sulfur-basalt concretes,*" Bureau of Mines Report No. RI 7349, U.S. Bureau of Mines, Washington, D.C., 21.

Czarnecki, B and Gillott. J.E. (1989) "Stress-strain behavior of sulphur concrete made with different aggregates and admixtures," *Quarterly Journal of Engineering Geology*, London, Vol. 22, pp. 195-206.

Dale, J.M. and Ludwig, A.C. (1968) "Advanced studies of sulfur aggregate mixtures as a structural material,' *Technical Report* No. AFWL-TR-68-21, Southwest Research Institute, San Antonio., 68.

Darnell, G. R., (1991) "Sulphur polymer cement, a New stabilization agent for mixed and low-level radioactive waste,' In *Proceedings of the First International Symposium on Mixed Waste,* Baltimore, Md., August 26-29, 1991, A. A. Moghissi and G. A. Benda, eds., Univ. of Maryland, Baltimore.

Duecker, W.W., (1934) "Admixtures improve properties of sulfur cements," *Chemical and Metallurgical Engineering*, Vol. 41, No. 11, pp. 583-586.

Funke, R.H.Jr., and McBee, W.C., (1982) "An industrial application of sulfur concretes," *ASC Symposium Series* No. 183, American Chemical Society, Washington, D.C., pp. 195-208.

Gemelli, E., Cruz, A.A.F., and Camargo, N.H.A. (2004) "A study of the application of residue from burned biomass in mortars," *Materials Research,* Vol. 7, No. 4, pp. 545-556.

Jordaan, I. J.; Gillott, J. E.; Loov, R. E., and Shrive, N. G., (1978) "Improved ductility of sulphur concretes and its relation to strength," *Proceedings of the International Conference on Sulphur in Construction*, Ottawa, (CANMET, Energy, Mines and Resources Canada), pp. 475-488.

Kalb, P. D., Heiser, J. H. and Colombo, P. (1991) "Modified sulfur cement encapsulation of mixed waste contaminated incinerator fly ash," *Waste Management,* Vol.11, No.3, pp.147-153.

Makenya, A.R. (2001) "Industrial application of sulfur concrete an environmental-friendly construction material," Dissertation, Dept. of Architecture, Royal Institute of Technology, Stockholm, Sweden, 232 pp.

Malhotra, V.M. (1974) "*Effect of specimen size on compressive strength of sulfur concrete,*' Division Report No. IR 74-25, Energy, Mines and Resources Canada Ottawa.

Malhotra, V.M., Soles, J. A., Carette, G. G. (1978) "*Stability of Sulfur-Infiltrated Concrete in Various Environments.*" Canada Centre for Mineral and Energy Technology, Department of Energy, Mines and Resources, Ottawa, Canada.

McBee, W.C. and Sullivan, T.A. (1982) "*Modified Sulfur Cement,*" U.S. Patent No. 4, 311, 826.

McBee, W.C., and Sullivan, T.A., (1979) "*Development of specialized sulfur concretes,*' Bureau of Mines Report No. RI 8346, U.S. Bureau of Mines, Washington, D.C., 21.

McBee, W.C., Sullivan, T.A., and Jong, B.W. (1983) "*Corrosion-resistant sulfur concretes,*" Bureau of Mines Report No. 8758, U.S. Bureau of Mines, Washington, D.C., 28.

McBee, W.C., Sullivan, T.A. and Jong, B.W. (1981) "*Modified sulfur concrete for use in concretes, flexible paving, coatings, and grouts,*" Bureau of Mines Report No. RI 8545, U.S. Bureau of Mines, Washington, D.C., 24.

Mohamed, A.M.O. (2002) "Keynote Paper: Geoenvironmental aspects of chemically based ground improvement techniques," *4th International Conference on Ground Improvement Techniques,* 26-28 March 2002, Kuala Lumpur, Malaysia.

Mohamed, A.M.O. (2003) "Geoenvironmental aspects of chemically based ground improvement techniques for pyritic mine tailings," *Ground Improvement,* Vol. 7, No. 2, pp. 73-85.

Mohamed, A.M.O., and El Gamal, M. (2006) "Compositional control on sulfur polymer concrete production for public works," In: *Developments in Arid Regions Research,* Vol. 3, pp. 27-38.

Mohamed, A.M.O. and El Gamal, M. (2007a) "Durability and leachability characteristics of modified sulfur cement and concrete barriers for containment of hazardous waste in arid lands," 1st Joint QP-JCCP

Environment Symposium in Qatar, "*Sustainable Development and Climate Change*", February 5-7, 2007, Doha, Qatar.

Mohamed, A.M.O. and El Gamal, M. (2007b) "Development of modified sulfur cement and concrete barriers for containment of hazardous waste in arid lands," *Sustainable Development and Climate Change*", February 5-7, 2007, Doha, Qatar.

Mohamed, A.M.O. and El-Gamal, M. (2007c) "Sulfur based hazardous Waste Solidification," *Environmental Geology*, Vol. 53, No. 1, pp. 159-175.

Mohamed, A.M.O. and El Gamal (2008a) "Sulfur cement and concrete production," *Proceedings of The 9th UAE University Annual Conference*, Al Ain, April 23-25, 2008, pp. ENG-1-10

Mohamed, A.M.O. and El Gamal, M. (2008b) "*New use of surfactant*" UK Patent Application No. 0807612.7, filed by J.A. KEMP and Co., UK, dated 25 April 2008.

Mohamed, A.M.O., and El Gamal, M. (2009a) "Hydro-mechanical behavior of a newly developed sulfur polymer concrete," *Cement and Concrete Composites,* Vol. 31, pp. 186-194.

Mohamed, A.M.O. and El Gamal, M. (2009b) "New use of surfactant" International Patent Application No. PCT/IB2009/005338, filed by J.A. KEMP and Co., UK, dated 21 April 2009.

Mohamed, A.M.O. and El Gamal, M. (2009c) "New use of surfactant" GCC Patent Application No. GCC/P/2009/13350, filed by J.A. KEMP and Co., UK, dated 25 April 2009.

Mohamed, A.M.O., and El Gamal, M. (2010) "Sulfur concrete for the construction industry: A sustainable development approach," *J. Ross Publishing,* Florida, USA, 424p.

Mohamed, A.M.O., Hossein, M., and Hassani, F.P. (2002) "Hydro-mechanical evaluation of stabilized mine tailings," *Environmental Geology Journal*, Vol. 41, pp. 749-759.

Mohamed, A.M.O., Hossein, M., and Hassani, F.P. (2003) "Role of fly ash addition on ettringite formation in lime-remediated mine tailings," *Journal of Cement, Concrete and Aggregates,* ASTM, December 2003, Vol. 25, No. 2, pp. 49-58 .

Sullivan, T.A. (1986) "*Corrosion-resistant sulfur concretes – design manual,*" The Sulphur Institute, Washington D.C., 44 .

Sullivan, T.A.; Mc Bee, W.C., and Blue, D.D. (1975) "Sulfur in coatings and structural materials," *Advances in Chemistry Series* No. 140, American Chemical Society, Washington, D.C., 55-74.

Vroom, A. H., (1977) "*Sulfur cements, process for making same and sulfur concretes made thereform*," U.S. Patent No. 4,058,500.

Vroom, A.H. (1981) "*Sulfur cements process for making same and sulfur concretes made there form*." U.S. Patent No. 4,293,463.

Vroom, A.H. (1998) "Sulfur concrete goes global," *Concrete International*, Vol. 20, No.1, pp. 68-71.

Vroom, A.H., (1992) "Sulfur polymer concrete and its application," In. *Proceeding of Seventh International Congress on polymers in concrete* Moscow: (1992) 606-621 .

Warren, C.J. and Dudas, M.J. (1992) "Acidification adjacent to an elemental sulfur stockpile: I. Mineral weathering," *Canadian Journal of Soil Science*, Vol. 72, pp.113-126.

Weber, H. H., McBee, W. C., Krabbe, E.A. (1990) 'Sulfur concrete composite materials for construction and maintenance," *Materials Performance*, Vol. 29, No. 12, Dec, 1990, pp. 73-77.

Wrzesinski, W.R., and McBee, W.C. (1988) "*Permeability and corrosion resistance of reinforced sulfur concrete,*" Report to the U.S. Bureau of Mines, Pittsburgh, PA. U.S. Government Printing Office: 1988 –605-017/80.001.13pp.

Young, R.N., Mohamed, A.M.O., and Warkentin, B.P., (1992) "*Principles of contaminant transport in soils,*" Elsevier, Amsterdam, the Netherlands, 327p.

In: Sand Dunes
Editor: Jessica A. Murphy, pp. 75-93
ISBN 978-1-61324-108-0

Chapter 3

SPIDER AS A MODEL TOWARDS THE CONSERVATION OF COASTAL SAND DUNES IN URUGUAY

Anita Aisenberg*[*1], Miguel Simó[2] and Carolina Jorge[2]
[1]Laboratorio de Etología, Ecología y Evolución,
Instituto de Investigaciones Biológicas Clemente Estable,
Avenida Italia 3318, Montevideo, Uruguay
[2]Sección Entomología, Facultad de Ciencias, Universidad de la República,
Iguá 4225, Montevideo, Uruguay.

ABSTRACT

Allocosa brasiliensis is a wolf spider that constructs burrows along the coastal sand dunes of Uruguay. They are medium-sized nocturnal spiders with whitish coloration that turns them cryptic with their sandy habitat. The species shows a reversal in both sex roles and sexual size dimorphism expected for spiders, being the first case reported in Araneae. Females are the mobile sex that looks for sexual partners and initiates courtship, and males are larger than females. Females prefer to mate with males showing deeper burrows. Copulations occur inside the male burrows and, after copulation ends, males donate their own burrows to the females, which officiate as nests for the future progeny. The ability of digging deep burrows in the sand is an adaptation to survive in the harsh

* E-mail: aisenber@iibce.edu.uy

coastal ecosystem where these spiders face drastic temperature variations, excessive heats, strong winds and prevent water loss. The atypical sex strategies found in this wolf spider could be related to ecological constraints. Prey availability fluctuates, showing high variability in quantity and quality, and high levels of cannibalism were observed. During the last decades, the Uruguayan coastline has been dramatically reduced and modified due to the introduction of exotic flora, urbanization and tourism, what has lead to habitat reduction and isolation of *A. brasiliensis* populations. The strict association between this spider and the sand dunes, in addition to its vulnerability to human impact and role as a generalist carnivore, turn this wolf spider into a good candidate for terrestrial biological indicator of conservation in coastal ecosystems. Behavioral, evolutionary and ecological studies on this species are essential for implementing adequate management and conservation plans for these areas.

INTRODUCTION

Lycosids, also named as wolf spiders, are small to large wandering spiders that have their body covered with hair and show a characteristic eye disposition: four small frontal eyes and four large posterior eyes (Foelix 1996). Secondary eyes of wolf spiders possess a membrane called tapetum that reflects external light and makes them easily recognizable when we light them with our head-lamps during the night, still from considerable distances. Female wolf spiders are also known for their brood care (Foelix 1996) (Figure 1). They carry their egg-sacs attached to their spinnerets and spiderlings climb to their mothers' dorsum after emergence. They stay on the mother's back until they are ready to disperse. Lycosids are distributed around the world and some species are synantropic. Very frequently we can find them in our gardens and, more rarely, inside our own houses.

Allocosa brasiliensis (Petrunkevitch 1910) is called white sand spider or sand dune wolf spider. Individuals construct burrows along the coastal sand dunes of Uruguay, Argentine and Brazil (Capocasale 1990). They are nocturnal spiders of approximately 2 cm of body length and whitish coloration, what turns them cryptic with the sandy habitat (Costa 1995) (Figure 2). Their reproductive period takes place during the summer of the Southern hemisphere, from November to April (Costa et al. 2006). At that time and during the night the spiders can be found walking on the sand dunes capturing preys or looking for potential sexual partners. As most spiders, they are generalist predators that prey on other invertebrate arthropods inhabiting the coastline.

Figure 1. Female of *A. brasiliensis* with her egg-sac attached to the spinnerets (Photograph: Authors).

Figure 2. Adult male of *A. brasiliensis* (Photograph: Marcelo Casacuberta).

This wolf spider shows a reversal in typical sex roles and sexual size dimorphism expected for spiders, being the fist case of sex role reversal cited for spiders (Aisenberg et al. 2007; Aisenberg and Costa 2008). These ethological particularities require morphological adaptations of each sex to their roles and could be a consequence of the drastic environment where the species inhabits. Additionally, the strict association between *A. brasiliensis* and coastal sand dunes, and the dramatic reduction of the coastal ecosystems in Uruguay due to urbanization makes them potential terrestrial bio-indicators of anthropogenic stress in these areas (Simó et al. 2005; Aisenberg et al. 2009). In the following sections we will overview aspects of the ethology, ecology and conservation of the species.

THE URUGUAYAN COASTLINE

The Uruguayan coastline expands along 714 km, about the Río de la Plata River and the Atlantic Ocean. The Río de la Plata River is a big estuary that receives fresh water from Uruguay River and Paraná River, and salty water from the Brazilian current of the Atlantic Ocean. The geologic formation of the Uruguayan coast is constituted by ancient rocks from the Paleoproterozoic (2300 MA) and Cambric (500 MA), with Cenozoic sedimentary rocks and Recent Quaternary sediments (Goso and Muzio 2006). The landscape is characterized by microtidal coasts with beaches of variable width that go from dissipative to reflective, with litoral drifts mainly directed to the South-West or West, excepting at the Eastern coast where they direct to the North-East. In winter, the dominant winds blow from the South-West and in summer from the South-East and North East (Panario and Gutiérrez 2006).

The granulometry of sediments deposited along the coastline go from thin to thick. Although the coast is a continuum, its morphology and associated landscapes differ along the coastline and comprises wide and long beaches, rocky points, sedimentary gullies, sand dunes, littoral lakes and wetlands (Gómez-Pivel 2006; Gómez and Martino 2008) (Figure 3). Two types of sand dunes are recognized. The first is represented by lines of dunes that stand near the shore and are shaped by water and wind. The sand of this type of dune is fixed by native plants as *Blutaparon portulacoides* (Amaranthaceae), *Panicum racemosum* and *Spartina ciliata* (Poaceae), *Cakile maritima* (Brassicaceae), *Hydrocotyle bonariensis* and *Senecio crassiflorus* (Apiaceae) and *Androtrichium trigynum* (Cyperaceae) (Costa 1995; Alonso-Paz and Bassagoda 2006) (Figure 4).

Figure 3. Landscapes of the Uruguayan coastline: a) fixed sand dunes (Photograph: Authors); b) dynamic sand dunes (Photograph: Daniel Panario and Ofelia Gutiérrez); c) rocky point (Photograph: Marcelo Casacuberta); d) gully (Photograph: Authors).

Figure 4. Typical vegetation of fixed sand dunes (Photograph: Authors).

The coastline also shows dynamic dunes that during the last decades have been strongly reduced to small areas on the Atlantic coast of the country. Many decades ago these sand dunes reached 30 meters of altitude. They were molded by the reduction of the sea level during the Quaternary and wind action (De Alava and Panario 1996). Originally, psamofile forests were expanded along the coast and between the water courses but today only relicts of them survive at the Southern coast (Ríos et al. 2010). Bushes contribute in the first steps of the fixation process of the dunes and the development of the substrate for the forests. This ecosystem contains a great diversity of flora that comprises about 1000 phanerogam species and twelve endemic taxa. Some xerofit species as *Cereus uruguayanus* R. Kiesling or species of the genus *Opuntia* (Cactaceae) are common in these areas (Alonso-Paz and Bassagoda 2006).

Burrow Digging in the Sand Dunes

Animals inhabiting the sandy coastline of Uruguay show adaptations to face the extreme heats during the day and cold temperatures during the night, prevent water loss and avoid the strong winds that are frequent in these areas (Costa et al. 2006; Aisenberg et al. 2009). One effective way for facing these conditions is to bury in the sand during daylight and restrict most activities to the night hours. In the case of spiders, burrow digging in the sand has been reported as an energetically expensive activity (Henschel and Lubin 1992). This could be due to three main factors: a) the cost of the production and deposition of silk necessary for avoiding burrow collapse; b) digging activities per se; 3) predation risk for the animals while they are exposed in the surface during burrow construction.

In the wolf spider *Allocosa brasiliensis* both natural and sexual selection would be driving efficiency in male burrow digging. Females do not construct long burrows but just short silk refuges where they stay during daylight. However, males have high selective pressures to construct deep and stable burrows. These burrows are preferred by females, they are the refuges used for copulation and will serve as nests for their progeny, because the female stays inside after mating (Aisenberg et al. 2007; Aisenberg and Costa 2008). Male burrows are approximately 10.0 cm long and 0.9 cm width, while female burrows are 3.0 cm of length and 0.8 cm width (Capocasale 1990; Aisenberg et al. 2007). Samplings at the field have also shown that *A. brasiliensis* adults of both sexes prefer to dig their burrows near the base of the dune (Aisenberg

2010; Aisenberg et al., in press) (Figure 5). The preference for this area of the dune could be related with higher prey availability, protection from the strong winds and high humidity levels that are reached closer to the surface. The burrows also are involved in temperature buffering, creating a more stable environment for eggs-sac care and spiderlings' development.

A. brasiliensis burrows can be located at distances of less than 10 cm among them (Aisenberg 2010). Males could be territorial and, after copulation, dig their new burrows close to the burrow with the copulated female. In this way, we could expect to find harems with one male and two or three females with their burrows clustered in the dune, hypotheses that require further testing. Individuals use their palps, front legs and chelicerae for sand extraction. While digging, they minimize their exposure outside the burrow and perform very fast movements and then return to the inside (Aisenberg and Peretti, unpublished data). Previous studies on this species found that males show a special feature with the shape of a spade at the tip of the palpal tarsi, instead of the typical claw of juveniles and females (Aisenberg et al. 2010).

This type of hoof probably aids in male digging. Also, we found that the male palpal organ is located in a more proximal position in comparison with other lycosids, what could be interpreted as a protection for this organ during excavation (Aisenberg et al. 2010).

Figure 5. Individual near a burrow entrance (Photograph: Marcelo Casacuberta).

The digging behaviors that show the longest durations are those related with silk deposition on the walls and around the burrow entrance (Aisenberg and Peretti, unpublished data). Individuals need to deposit multiple layers of silk for constructing a stable burrow in the open dunes. As silk production is costly and, in the case of males they need to construct deep burrows, we can expect digging behavior to be energetically demanding for these wolf spiders.

In general, burrow entrances are closed with silk and sand during the day and opened during the night. Closing the entrance ensures a better camouflage of the burrow, complicating their detection by the diurnal *Anoplius* pompilid wasps, common in the same areas than *A. brasiliensis* and parasitoids of this species (Costa 1995).

Sex Role Reversal in a Wolf Spider Inhabitant of Coastal Sand Dunes

Sex roles reflect female and male contributions in production of gametes, courtship and copulatory effort, delivery of nuptial gifts, or other resources associated with reproduction and parental care (Trivers 1972; Bonduriansky 2001). Pre-copulatory asymmetries between the sexes in gamete production will promote asymmetries in investment during and after copulation. As a consequence, in general, males that invest less in gamete production will try to maximize the number of copulations and they will be the roving sex that looks for potential sexual partners, competes for access to females and displays elaborate courtships (Andersson 1994). On the other hand, females will be the selective sex that has a higher investment in parental care. However, there are exceptions to this rule.

In species with high male reproductive investment, typical sex roles can reverse from their expected patterns and males can turn choosy while females compete for access to males and initiate courtship (Gwynne 1991; Andersson 1994; Bonduriansky 2001). Sex role reversal has been cited for insects, crustaceans, fish, amphibians and birds (Gwynne 1991; Andersson 1994; Eens and Pinxten 2000; Bonduriansky 2001).

Harsh environments with high fluctuations in prey abundance and rough environmental conditions have been indicated as potential settings for cases of sex role reversal (Karlsson et al. 1997; Lorch 2002). Although sex role reversal has been cited in various animal taxa, few cases have been thoroughly studied. Many authors have highlighted the importance of studying these atypical cases towards a more robust theory of sex roles and factors determining them (Eberhard 1985; Tallamy 2000; Bonduriansky 2001; Roughgarden et al. 2006).

Allocosa brasiliensis shows a reversal in typical sex roles and expected sexual size dimorphism (Aisenberg et al. 2007). Males are larger than females, in opposition to what is the norm in spiders, and females are the roving sex that goes out and locates male burrows and initiates courtship (Aisenberg et al. 2007). Females prefer to mate with those males showing the longest burrows, but males are also choosy and prefer virgin females with good body condition. Males are so dramatically selective that when they do not accept a female for copulation they can cannibalize her (Aisenberg et al. 2011). If copulation takes place, it occurs inside the male burrow and after it ends, the male donates his burrow to the female. Both sexes collaborate in closing the burrow entrance, the male from the outside and the female from inside the burrow. The male will need to construct another burrow to access to new matings. So, why are males so devoted and females so daring in this system? Individuals of *Allocosa* genus are the only wolf spiders that are adapted to inhabit in the Uruguayan coastal sand dunes, where refuges and vegetation are scarce, and prey abundance is variable and highly dependant of the climatic conditions (Celentano and Defeo 2006; Costa et al. 2006; Aisenberg et al. 2009).

Figure 6. Adult female carrying the spiderlings on her dorsum (Photograph: Authors).

Female wolf spiders forage intensively before copulation but do not feed during egg-sac care and until the emergence of the spiderlings (Figure 6) (Capocasale and Costa 1975; Wagner 1995). So, females of *A. brasiliensis* need to forage in a habitat where preys can be scarce for long periods and they also need to find a safe refuge where they can lay their eggs. Strong pressures related with the environment and the condition of wolf spider could be driving male high reproductive investment as supplier of the mating refuge and nest of the future progeny.

To Eat a Relative or Not to Eat

Allocosa brasiliensis is a generalistic and opportunistic predator. Preys in the sand dunes can be scarce and variable, so this spider is able to vary the diet according to availability. The most frequent preys are spiders, flies and ants (both workers and winged individuals). They also prey on Diptera, Hymenoptera, Coleoptera, Lepidoptera, Homoptera and Orthoptera (Aisenberg et al. 2009; Figure 7). Ants are very abundant in the sand dunes, especially during the summer (Costa et al. 2006) and we can say they are the most predictable preys of this habitat. The consumption of ants by other wolf spiders species can be considered rare. In this case, it suggests food limitation and the need of adaptations to manage these preys with mechanical and chemical defenses.

A. brasiliensis shows high rates of intra-guild predation. This means that it is frequent to find them preying on other wolf spiders (Aisenberg et al. 2009), including individuals from their own species and always following the rule: the big eats the small. Moreover, data from samplings at the field indicate that adult males can attack and cannibalize adult females from their own species (Aisenberg et al. 2009). Cannibalism is widespread among wolf spiders (Fernández-Montraveta and Ortega 1990; Wagner and Wise 1996; Moya-Laraño et al. 2002; Wise 2006), but male cannibalism on females is extremely rare, not only for spiders but for the animal kingdom in general (Elgar 1998; Elgar and Schneider 2004; Wise 2006). Male cannibalism can be considered maladaptive in terms of losing a potential sexual partner. However, considering the high concentration of *A. brasiliensis* individuals in some of these areas and that mated females remain buried there until spiderling dispersal (Aisenberg et al. 2007), a walking female could turn into a good meal for a hungry male. In addition, males of *A. brasiliensis* are larger and show bigger and more sclerotized chelicerae compared to females, what offers obvious advantages to males in terms of suffering injuries when they perform their attacks over the members of the other sex.

Figure 7. Individual of *A. brasiliensis* preying on a larva (Photograph: Authors).

Human Impact

Since the beginning of the twentieth century, the Uruguayan Southern coast has been modified by human impact, mainly for the development of urbanization areas, fixation of mobile dunes with exotic trees and tourism activities. These changes promoted a strong modification in the landscapes that were originally occupied by several kilometers of sand dunes and were later reduced to one or two hundred meters in some sites. Changes in the original vegetation of these areas also promoted the invasion of exotic animals that compete with native fauna.

Today, along the coast of Uruguay we can recognize a first zone located 60 m from the shore that contains the dune line composed by dry sand and a second zone behind, localized at approximately 80 m from the shore, that is characterized by the presence of dune lines with psamophile vegetation. Behind these two areas, grows the exotic *Acacia longifolia* Wild (Leguminosae), a common shrub in the coast that dominates the landscape over the native flora (Alonso-Paz and Bassagoda 2006). Further away from the shore, the area is characterized by the presence of exotic tree species as *Pinus* spp. and *Eucalyptus* spp (Figure 8).

Figure 8. Exotic vegetation on the Uruguayan coast (Photograph: Authors).

The highest urbanization and the most important tourism activities take place at the Southern coast of Uruguay. This has provoked a strong process of fragmentation occurring in many beaches, reducing and isolating them and their corresponding biological corridors. Fragmentation affects the normal dispersal of the individuals and prevents or diminishes gene flow between natural populations. As an example, *Allocosa brasiliensis* and *Allocosa alticeps* are two wolf spider species that inhabit the sandy line coast. Today, the reduction, modification by plantation and fragmentation of this habitat promoted local extinctions of these species.

Vagility and Distribution along the Sand Dunes

As was previously mentioned, the Uruguayan coastline has been reduced and notoriously fragmented (Costa 1995). Consequently, *A. brasiliensis,* as well as other species inhabitant of coastal areas, have suffered a significant habitat loss. For this reason, we carried studies focused on aspects of *A. brasiliensis* population dynamics on the Southern coast of Uruguay, with the objective of estimating the vagility of the species in areas with low and high human impact, analyzing the effects of habitat fragmentation. For that purpose, we captured, marked and recaptured individuals during nocturnal samples. This method has been successfully applied in other spider species to estimate population abundances (Kiss and Samu 2000). We observed 1195 individuals, 368 were marked and 18 were re-captured. Preliminary results indicate that the majority of individuals and nests would be on the first line of dunes from the shore, in areas with native psamophile, as had been proposed by Simó et al. (2005) and Aisenberg et al. (2009). We also confirmed that individuals of this species show heterogeneous spatial distribution, associated with the presence of sand dunes and native vegetation. Modified beaches showed a clear fall in the number of individuals and the highest numbers of recaptures, compared to beaches with lower levels of human impact. The average vagility of the individuals ranged between 0.1 and 115 meters, that was positively correlated with the width and total area of the beach considered.

Summarizing, processes of fragmentation and reduction of the coastal environment would be negatively affecting *A. brasiliensis* vagility and the viability of their populations, requiring our urgent attention towards the conservation of these areas.

The White Spider of the Sand Dunes as a Bio-Indicator

One taxon is considered bio-indicator when it supplies information about the ecosystem where it lives. This information is very important because it can be applied on the evaluation of the health of the environment considered. Spiders have been used as biological indicators to evaluate the anthropogenic stress caused by heavy metal pollution in natural habitats (Maelfait and Hendrickx 1998). Other studies with spiders focused on the evaluation of soil quality and conservation of natural habitats (Hartleyaet al. 2008; Uehara-Prado et al. 2009). These arthropods are useful models because they live in almost all terrestrial habitats and they can be easily studied, both at the field and under laboratory conditions.

Allocosa brasiliensis is specially adapted to inhabiting sandy substrates that suffer a great variation in temperature and humidity conditions. Previous studies showed that this species lives associated to open areas of sand dunes fixed by native psamophile vegetation. For these reason, *A. brasiliensis* was proposed as a good bio-indicator of the conservation along the Uruguayan coast. At present and as was cited before, several beaches of Uruguay have been fragmented by human constructions and introduction of exotic flora (Figure 9). Plans of restoration in some areas have proposed the installation of wood barriers that retain the sand (Panario and Gutiérrez 2006) (Figure 10).

Figure 9. Beach of the Rio de la Plata coast with high human impact (Photograph: Authors).

Figure 10. Wood barriers installed to help retain the sand dunes at a beach of Montevideo (Photograph: Authors).

This methodology could help maintain the natural habitat of this species, but it is not enough. Corridors between the beaches must be restored to allow gene flow between populations.

One interesting question is: how this species is involved in the colonization of the dunes by psamophile vegetation? According to field observations (Miguel Simó, unpublished data), seeds of native plants can be found inside the spider burrows, suggesting a correlation between the spider and native vegetation that represents an interesting fact to be further studied.

Conservation of the Uruguayan coastline should include a plan of substitution of exotic trees and shrubs by native flora and species of the original psamophile forest. The aim should be to restore the natural landscape, flora and fauna, while maintaining the touristic service that beaches offer. Nowadays, management plans for conservation in Uruguay only include the monitoring of aquatic microorganisms, but terrestrial bio-indicators should also be considered.

Conclusions and Future Perspectives

The loss of biodiversity due to direct or indirect human impact is one of the most extended environmental problems, that unfortunately is far from being solved. The distribution and abundance of many animal species is declining due to fragmentation of their habitats that determine, in many cases, the viability of the populations and increase the likelihood of local extinctions.

Several threats affect the conservation of the Uruguayan coast. The forestation with exotic plants that occurred at the beginning of the twentieth century produced modifications in the profile of the coast, increasing the erosion by the sea and promoting dune crumble. The 70% of Uruguayan human population is established near the Río de la Plata coast. The construction of roads and buildings near the coast has also affected the structure and dynamics of the dunes. The extraction of sand for constructions has diminished the dune line, generating in many cases artificial lakes.

The establishment of conservation plans on the landscape of the Uruguayan Southern coast is urgently needed to preserve the coastal dynamics and the native flora and fauna. As well as an interesting evolutionary and ethological model, *A. brasiliensis* postulates as a promising bio-indicator of human impact on the Uruguayan coast, where the presence, abundance, shortage or absence of this species reflects the degree of conservation on these areas.

BIBLIOGRAPHY

Aisenberg A. 2010. Causas y consecuencias de la inversión de roles sexuales y dimorfismo sexual invertido en *Allocosa alticeps* y *Allocosa brasiliensis* (Araneae, Lycosidae), dos especies nativas de los arenales costeros. PhD thesis, PEDECIBA, Universidad de la República, Montevideo, Uruguay.

Aisenberg A and Costa FG. 2008. Reproductive isolation and sex role reversal in two sympatric sand-dwelling wolf spiders of the genus *Allocosa. Canadian Journal of Zoology*. 86:648-658.

Aisenberg A, Costa FG and González M. 2011. Male sexual cannibalism in a sand-dwelling wolf spider with sex role reversal. Biological Journal of The Linnean Society. D.O.I. 10.1111/j.1095-8312.2011.01631.x

Aisenberg A, Costa FG, González M, Postiglioni R and Pérez-Miles F. 2010. Morphological and morphometrical adaptations for reversed sex roles in two sand dune wolf spiders. *Journal of Natural History* 44:1189–1202.

Aisenberg A, González M, Laborda A, Postiglioni R and Simó M. 2009. Reversed cannibalism, foraging, and surface activities of *Allocosa alticeps* and *Allocosa brasiliensis*: two wolf spiders from coastal sand dunes. *Journal of Arachnology*. 37(2):135-138.

Aisenberg A, González M, Laborda Á, Postiglioni R and Simó M, in press. Spatial distribution, burrow depth and temperature: implications for the sexual strategies in two *Allocosa* wolf spiders. Studies of Neotropical Fauna and Environment

Aisenberg A, Viera C and Costa FG. 2007. Daring females, devoted males and reversed sexual size dimorphism in the sand-dwelling spider *Allocosa brasiliensis* (Araneae, Lycosidae). *Behavioral Ecology and Sociobiology* 62:29-35.

Alonso-Paz E and Bassagoda MJ. 2006. Flora y vegetación de la costa platense atlántica. In: *Bases para la conservación y el manejo de la costa uruguaya* (Menafra R, Rodríguez-Gallego L, Scarabino F and Conde D, eds.). Vida Silvestre, Uruguay. Pp. 21-34.

Andersson M.1994. *Sexual Selection*. Princeton University Press, Princeton, New Jersey.

Bonduriansky R. 2001. The evolution of male mate choice in insects: a synthesis of ideas and evidence. *Biological Reviews* 76:305-339.

Capocasale RM.1990. Las especies de la subfamilia Hippasinae de América del Sur Araneae, Lycosidae). *Journal of Arachnology* 18:131-134.

Capocasale RM and Costa FG (1975) Descripción de los biotopos y caracterización de los habitats de *Lycosa malitiosa* Tullgren (Araneae: Lycosidae) en Uruguay. Vie Milieu Sér C Biol Terr 25(1):1–15.

Celentano E and Defeo O. 2006. Habitat harshness and morphodynamics: life history traits of the mole crab *Emerita brasiliensis* in Uruguayan sandy beaches. *Marine Biology* 149:1453-1461.

Costa FG. 1995. Ecología y actividad diaria de las arañas de la arena *Allocosa* spp (Araneae, Lycosidae) en Marindia, localidad costera del Sur del Uruguay. *Revista Brasileira de Biologia* 55(3):457-466.

Costa FG, Simó M and Aisenberg A. 2006. Composición y ecología de la fauna epigea de Marindia (Canelones, Uruguay) con especial énfasis en las arañas: estudio de dos años con trampas de intercepción. In: *Bases para la Conservación y el Manejo de la Costa Uruguaya* (Menafra R, Rodríguez-Gallego L, Scarabino F and Conde D, eds.). Vida Silvestre, Uruguay. Pp. 427-436.

De Álava D. and Panario D. 1996. La Costa Atlántica: ecosistemas perdidos y el nacimiento de un monte de pinos y acacias. In: *Almanaque Banco de Seguros del Estado*. Montevideo.

Eberhard WG. 1985. Sexual Selection and animal genitalia. Harvard University Press, Cambridge, Massachusetts.

Eens M. and Pinxten R. 2000. Sex-role reversal in vertebrates: behavioural and endocrinal accounts. *Behavioural Processes* 51:135-147.

Elgar MA. 1998. Sperm competition and sexual selection in spiders and other arachnids. In: *Sperm competition and Sexual Selection* (Birkhead TR and Moller AP, eds.). Academic Press, London. Pp. 307-339.

Elgar MA and Schneider JM. 2004. Evolutionary significance of sexual cannibalism. *Advances in the study of behavior* 34:135-163.

Fernández-Montraveta C and Ortega J. 1990. El comportamiento agonístico de hembras adultas de *Lycosa tarentula fasciiventris* (Araneae, Lycosidae). *Journal of Arachnology* 18:49-58.

Foelix RF. 1996. *Biology of spiders*. Oxford University Press, New York.

Gómez-Pivel MA. 2006. Geomorfología y procesos erosivos en la costa atlántica uruguaya. In: *Bases para la Conservación y el Manejo de la Costa Uruguaya* (Menafra R, Rodríguez-Gallego L, Scarabino F and Conde D, eds.). Vida Silvestre, Uruguay. Pp. 35-43.

Gómez M and Martino D. 2008. *GEO Uruguay, Informe del Estado del Ambiente*. Montevideo, CLAES/PNUMA/DINAMA.

Goso CA and Muzio R. 2006. Geología de la costa uruguaya y sus recursos minerales asociados. In: *Bases para la Conservación y el Manejo de la*

Costa Uruguaya (Menafra R, Rodríguez-Gallego L, Scarabino F and Conde D, eds.). Vida Silvestre, Uruguay. Pp. 9-19.

Gwynne DT. 1991. Sexual competition among females: What causes courtship-role reversal? *Trends in Ecology and Evolution* 6(4):118-121.

Hartleya W, Uffindella L, Plumba A, Rawlinsonb HA, Putwainc P and Dickinson NM. 2008. Assessing biological indicators for remediated anthropogenic urban soils. *Science of the Total Environment* 405:358–369.

Henschel JR, and Lubin YD. 1992. Environmental factors affecting the web and activity of a psammophilous spider in the Namib Desert. *Journal of Arid Environments* 22:173-189.

Karlsson B, Leimar O and Wiklund C. 1997. Unpredictable environments, nuptial gifts and the evolution of size dimorphism in insects: an experiment. *Proceedings of the Royal Society of London* 264:475-479.

Kiss B and Samu F. 2000. Evaluation of populations density of the common wolf spider Pardosa agrestis (Araneae; Lycosidae) in Hungarian alfalfa fields using mark-recapture. *European Journal of Entomology*. 97:191-195.

Lorch P. 2002. Why is mutual mate choice not the norm? Operational sex ratios, sex roles and the evolution of sexually dimorphic and monomorphic signalling. *Philosophical Transactions of the Royal Society of London B* 357:319-330.

Maelfait JP and Hendrickx F. 1998. Spiders as bio-indicators of anthropogenic stress in natural and semi-natural habitats in Flanders (Belgium): some recent developments P. A. Selden (ed.). *Proceedings of the 17th European Colloquium of Arachnology*, Edinburgh. 293-300.

Moya-Laraño J, Orta-Ocaña JM, Barrientos JA, Bach C and Wise DH. 2002. Territoriality in a cannibalistic burrowing wolf spider. *Ecology* 83(2):356-361.

Panario D and Gutiérrez O. 2006. Dinámica y fuentes de sedimentos de las playas uruguayas. In: Bases para la conservación y el manejo de la costa uruguaya. (Menafra R, Rodríguez-Gallego L, Scarabino F and Conde D, eds.). Vida Silvestre, Uruguay. Pp. 21-34.

Ríos M, Bartesaghi L, Piñeyro V, Garay A, Mai P, Delfino L, Masciadri S, Alonso Paz E, Bassagoda M and Soutullo A. 2010. Caracterización y distribución espacial del bosque y matorral psamófilo. EcoPlata, SNAP, Serie de Informes N23, julio. Available: http://www.guayabira.org.uy/psamofilo/Psamofilo_Ecoplata_2010

Roughgarden J, Oishi M and Akcay E. 2006. Reproductive social behavior: cooperative games to replace sexual selection. *Science* 311:965-969.

Simó M, Coll JJ, Viglioni M and Laborda A. 2005. Actividad, distribución espacial y fragmentación del hábitat en dos especies de *Allocosa* Banks, de la costa sur del Uruguay (Araneae, Lycosidae). In Actas Primer Congreso Latinoamericano de Aracnología, Minas, Uruguay: 148.

Tallamy DW. 2000. Sexual selection and the evolution of exclusive paternal care in arthropods. *Animal Behaviour* 60:559-567.

Trivers RL. 1972. Parental investment and sexual selection. In: *Sexual selection and the descent of man 1871-1971* (Campbell B, ed.). Aldine, Chicago. Pp: 136-179.

Uehara-Prado M, de Oliveira Fernandes J, de Moura Bello A, Machado G, Santos A, Zagury Vaz-de-Mello F and Lucci Freitas A. 2009. Selecting terrestrial arthropods as indicators of small-scale disturbance: A first approach in the Brazilian Atlantic Forest. *Biological Conservation* 142(6):1220-1228.

Wagner JD. 1995. Egg sac inhibits filial cannibalism in the wolf spider *Schizocosa ocreata. Animal Behaviour* 50:555–557.

Wagner JD and Wise DH. 1996. Cannibalism regulates densities of young wolf spiders: evidence from field and laboratory experiments. *Ecology* 77:639-652.

Wise DH. 2006. Cannibalism, food limitation, intraspecific competition, and the regulation of spider copulations. *Annual Review of Entomology* 51:441-465.

Reviewed by: Fernando G. Costa.

In: Sand Dunes ISBN 978-1-61324-108-0
Editor: Jessica A. Murphy, pp. 95-117

Chapter 4

OCCURENCE OF ACTINOMYCETES AND MYCORRHIZAE ASSOCIATED WITH NATURAL AND RECONSTRUCTED SAND DUNE VEGETATION ZONES DISTURBED BY PREVIOUS SAND-MINING OPERATIONS ON AUSTRALIA'S FRASER ISLAND

D. İ. Kurtböke and S. E. Bellgard*
Faculty of Science, Health and Education,
University of the Sunshine Coast,
Maroochydore DC, QLD 4558, Australia

ABSTRACT

The eastern coastline of Australia became subjected to intensive mining activity in the 19th and 20th centuries, which has resulted in disturbed sand dune systems. Public pressure has subsequently led to some protection measures and restoration programs in disturbed sand dune areas began to be implemented with varying degrees of success The re-modelled surficial landscape resembles the surrounding landscape in

* Current address: Biodiversity and Conservation), Landcare Research, Private Bag 92170, Auckland Mail Centre, Auckland 1142, New Zealand

that local species have been planted to increase the success rate. Mine site rehabilitation on Fraser Island, now a World Heritage Site, however, has produced poor biodiversity-outcomes over time. This is reflected in species-poor post-disturbance plant communities that lack key-stone species and any structural heterogeneity to provide critical wildlife habitat. Understanding the occurrence, distribution and the role of the soil-biotic factors involved in the natural recovery process may therefore benefit the revegetation and restoration of disturbed sand dune systems. The chapter presented here questions whether disturbance of healthy actinomycete and mycorrhizal populations in the course of sand mining, topsoil stripping, storage and re-application is among the reasons for poor recovery response observed on the island over the last 30-40 years.

Keywords: Actinomycetes, Mycorrhizae, Sand dunes, Sand mining.

Fraser Island Sand Dunes

The Australian state of Queensland offers some of the best-known examples of coastal sand dune and vegetation systems worldwide, which have survived partly due to their size and inaccessibility. Two such examples are the sand dunes of Fraser and Moreton Islands showing the extent to which coastal vegetation can develop. For hundreds of thousands of years complex forest communities have grown on sand credit to the simple assemblage of grasses, herbs and shrubs on these islands (Greening Australia Queensland (Inc.) 2001).

Fraser Island, which covers an area of 184000 hectares, is the world's largest sand island. It is located in Queensland and displays an exceptionally diverse biome ranging from coastal dune vegetation through to forest dominated by Queensland kauri (*Agathis robusta*). The Island is 123 km long with an average width of 14km with sand dunes of transgressive, parabolic quartz believed to vary in age from present to 700,000 years (http://www.unep-wcmc.org/sites/wh/pdf/Fraser%20Island.pdf).

Pioneer species that grow along foredunes include salt tolerant grasses, creepers and groundcovers such as beach-spinifex (*Spinifex sericeus*), pig-face (*Carpobrotus* spp.), goats-foot (*Ipomaea pes-caprae*) and guinea-flower (*Hibbertia scandens*) (Greening Australia Queensland (Inc.) 2001). A distinctively taller band is formed behind the pioneer species by coastal-sheoaks (*Casuarina equisetifolia*) pruned on the seaward side due to salt-laden winds. Other vegetation communities such as Banksia woodlands, mangroves, salt marsh and Melaleuca wetlands develop further inland (Jehne and Thompson 1981). Fraser Island's extensive dune systems support more mature, complex communities including rainforest and wet sclerophyll forest. Today, in the natural areas, the variety of vegetation still remains extraordinary, ranging from Coastal Heath to Wallum, Open Woodlands, Mangrove and Rainforest.

Dune systems are the building blocks of Fraser Island. The sand on the Island is the result of erosion of the Great Dividing Range along the East Coast of Australia. A continental drift pattern pushes this sand up onto the coast of SE Queensland (Clutterbuck 1977). As sand is deposited onto the east coast of the island, it is blown up the beach and creates the first dunes called foredunes. Generally, vegetation such as spinifex grasses stabilize the sand and prevent it from moving inland (Figure 1). Foredunes lead to successional dunes finally ending up with mature hind dunes (http://www.environment.gov.au/heritage/places/world/fraser/resources.html).

However, disturbances in the vegetation at times might push the sand across the island by the wind forming a "U shaped" dune system called a Parabolic dune. Parabolic dunes move, often slowly, in giant "sand waves" across the island covering over older dune systems that may have been previously stabilized by vegetation. It is known that dead trees sticking out of the top of the sand on these overtaking dunes can commonly be seen (http://www.tourfraserisland.com.au/sand_dunes.php). Natural disturbance has always been a normal part of beach and dune building processes which include (i) sand abrasion, (ii) erosion, (iii) build up of sand, (iv) wind blasting and dehydration, (v) wave attack. These influences may increase in intensity during storm events and king tides (Greening Australia Queensland (Inc.) 2001).

Figure 1. (a) Overall view of stretching foredunes along the Island dominated by spinifex grasses.

Interestingly, some plants, such as *Spinifex* are able to survive, change their morphology and even flourish in response to changing conditions, such as *Spinifex*. Today, however, natural disturbances often threaten dune systems because of their reduced size and connectivity. Often, little or no natural buffer remains. Symptoms of "unhealthy" dune vegetation can be (i) changes in dune shape and stability, (ii) weed invasion and spread, (iii) severe erosion (e.g. foredune collapse, blowouts), (iv) loss of natural regeneration and recruitment of native species, (v) decline in local wildlife (Greening Australia Queensland (Inc.) 2001) (Figure 2a,b).

Figure 2. (a) Collapse of fore dunes and (b) blowouts due to poor coverage by canopy species.

SAND MINING ON THE ISLAND AND REHABILITATION EFFORTS

The natural coastal habitat of Fraser Island has been disturbed in the past due to the mining of mineral sand ilmenite (DEHWA 2010). The first area impacted by sand mining on the island occurred before 1973. The disturbed area is the hind dune south of Dilli Village in a linear strip (approx. 20 km long) back to North Spit, near the southern tip of the island (15^{0}09'28.79"E and 25^{0}14'07.27"S) (Figure 3).

While the initial strip mining activity commenced at the southern tip, there was also damage further north of the mining-front from the construction of seismic lines. These lines were etched into the hind dune as part of the pre-mining proving-up of the resource and disturbance to the hind dune system also occurred during this time. Throughout the mining years, large tracks of dunes were cleared and exposed for lengthy periods. Usually the top-soil (overburden) was removed and stockpiled, exposing the heavy mineral rich sands for mining and extraction by pond settling. During this period the sands remained un-vegetated and were subject to similar processes as natural sand blows.

Photo courtesy of John Sinclair, Go Bush Safaris, Australia's World heritage Specialists. http://www.sinclair.org.au/

Figure 3. Aerial view of the Dillingham Murphyores mining site circa 1975.

A process known as *succession* is reflected in the formation of distinct "zones" that run parallel to the shoreline. This process involves the succession of simple plant communities by a series of more complex communities early on showing how dynamic coastal vegetation communities are (Dobson *et al.* 1997).

The vegetation cover on the rehabilitated mine land is a structurally simplified plant community with reduced canopy cover, which has lower density of the canopy species that occur in the intact hind dune situation. The species composition of the rehabilitated area is comparable to the intact hind dune, however, there is predomination of prickly couch that forms a dense mat. This may be responsible for the reduced diversity of the ground cover/herbaceous layer in the rehabilitated dune community.

The obvious canopy species in the rehabilitated area are Coastal Sheoaks, but they are very spindly representatives, of similar age, lower density, providing little or no structural heterogeneity, and do little cover to the zone to serve as wildlife refuge (Figure 4).

Figure 4. View of rehabilitation site looking to fringing vegetation with thin crowns.

Role of Microorganisms in the Success of Revegetation of Sand Dunes and Fraser Island Findings

Dune vegetation is essential for the formation and preservation of sand dunes and protection of the coast lines (Rodriguez-Echeverria and Freitas 2006). Vegetated sand dunes differ from a pile of inert sand particles in many ways, but one of the most important characteristics they exhibit are their association with the populations of rhizosphere microorganisms (Maremmani *et al.* 2003, Park *et al.* 2005). Filamentous fungi are capable of binding particles of soil into stable aggregates because particles of soil adhere to a mucilage on the surface of hyphae (Tisdall and Oades 1979).

Coastal sand dunes are harsh environments with low nutrient content and especially lack nitrogen (N) and phosphorus (P). Arbuscular mycorrhizal fungi (AMF) play a vital role in the establishment and survival of dune colonizing plants, by aiding in nutrition and contributing to the process of dune stabilization by binding sand grains into wind resistant aggregates (Lemauviel and Roze 2003, Rodriguez-Echeverria and Freitas 2006). Furthermore, endophytic, symbiotic and nitrogen fixing bacteria also play a role in the maintenance of coastal sand dune plant health (Shin *et al.* 2007). Actinomycetes, particularly streptomycetes are frequently reported to produce antimicrobial and growth promoting compounds (Tan and Zou 2001), which might have importance in the early stages of the plant establishment. With the aid of such beneficial microflora producing wide range of plant beneficial compounds, pioneer plant species colonizing, stabilizing and thus conserving mobile coastal dunes might also be achieved. Furthermore, cellulase-producing micromonosporae reportedly protect *Banksia grandis* trees from *Phytophthora* infections when combined with streptomycete species producing streptomycete species (El-Tarabily *et al.* 1996). *Phytophthora* infections were reported in the dune plants of Australia (Hill *et al.* 1994, Shearer and Dillon 1996). Detection of actinomycetes in natural dunes might indicate that these organisms may be involved in the production of compounds that serve to protect plant roots from root-rot pathogens that can influence the success or failure of early development.

In Australia, an early assessment of the presence or absence of AMF in plants colonizing coastal sand-dunes at Cooloola revealed that *Pultenaea villosa* had endomycorrhizae (Jehne and Thompson (1981). These research studies also recovered the spores of the two species of AMF namely

Gigaspora calospora and *Glomus fasciculatus*. Peterson *et al.* (1985) examined 42 species of angiosperms from the all vegetation zones on Heron Island and 57% of the species had AMF associations.

Casuarinas are also important components of many forests and are major pioneers of low fertility sand dunes (Lubke 2008). Logan *et al.* (1989) found that 36 of the 41 sand–dune plant species they examined had AMF associations in the coastal NSW. In their study, *Casuarina equisetifolia* was observed to have both root nodules and typical AMF associations including arbuscules. Nitrogen fixation occurs in root nodules involving the actinomycete *Frankia casuarinae* (Becking 1977). Recently, it has been demonstrated that not only *Frankia* as a dinitrogen fixing symbiont, but also other saprophytic strains of actinomycetes have positive effects on the growth of actinorhizal plants (Solans *et al.* (2003). Additionally, previous studies in Senegal have identified that *Casuarina equisetifolia* can have three symbioses with soil microbes: 1) actinomycetaceous dinitrogen-fixing nodules, 2) arbuscular mycorrhizal fungi (AMF) in fine feeder roots, and 3) ectomychorrizal fungi sheathing the roots (Bâ *et al* 1987). Logan *et al.* (1989), Peterson *et al.* (1985), Logan *et al.* (1989), and Kurtböke *et al.* (2007) have found *C. equisetifolia* with the dual microbial symbioses of actinomycete N_2-fixing nodules and AMF. *Casuarina equisetifolia* represents one of the key-stone species of the mature hind dune and plays a role in long-term stability and for wildlife habitat through provision of cover, perches, coarse woody debris and nesting hollows.

Soil disturbance, like that associated with open-cut and surface-mining reduces the fertility and biological activity of the AMF (Bellgard 1992, Jasper *et al.* 1991). Brockoff and Allaway (1989) investigated AMF in natural vegetation and sand-mined dunes and found that the level of root length colonized by AMF was lower in mined soils (2-4 years). AMF has also been identified as a contributor to aggregate-formation in sand-dune soils (Sutton and Shepherd (1976). Consequently, the loss of these microbiological components (due to the deleterious effects of soil disturbance) could have an impact on the *in situ* soil pedogenesis that occurs after pioneer species colonise a sand dune.

Some studies indicate that while soil disturbance can reduce the infection inducing potential of soil, it does not necessarily destroy the AMF association completely (Gardner and Malajczuk 1988, Bellgard 1992, 1993, Jasper *et al.* 1992). The role mycorrhizae play in sand dune recolonization/succession depends upon the number of infection units/spores available to initiate infection with germinating seed after the mining has ceased. Furthermore, spores of AMF migrate both by wind and animals and insects (e.g. grasshoppers and crickets). Ants, earthworms and other soil invertebrates also play a role in the dispersal of whole sporocarps of AMF (Dell 2002; Klironomis and Moutoglis 2004). In strip mined areas, rodent dissemination of AMF fungal propagules has been shown to increase the rate of revegetation (Ponder *et al.* 1980; Rothwell and Holt 1978). Although AMF associations were found in the fore dunes of Fraser Island by the Authors, in particular with *Spinifex sericeus,* the lower levels of AMF-root colonization in *Casuarina* plants colonizing the rehabilitated dunes may be a reflection of the lower inoculum potential caused by the soil disturbance associated with the strip-mining process (Table 1).

Typical AMF structures were identified in the roots of *C. equisetifolia* growing in the rehabilitated dune of the Fraser Island indicating that the infectivity of the AMF propagules was not completely destroyed by the strip mining process (Table 1). Bellgard (1992) demonstrated that hyphae of mycorrhizal fungi grew from root fragments and spores after severe soil disturbance, and successfully initiated the mycorrhizal association in bioassay trap plants after 42-days.

Table 1. Mycorrhizal characteristics of fine roots from plants in three different dune zones at Fraser Island

Root sample	Length of root assessed	Root clusters in epidermal cells	Hyphal coils in cortical cells	Internal hyphae	Vesicles/ Arbu-scules	Range in % AMF infection
Spinifex from fore dune	18 cm	Absent	Present	Present	Present	90-95%
Casuarina from intact hind dune	91 cm	Present	Present	Present	Present	65-95
Casuarina from rehabilitated hind dune	72 cm	Present	Present	Present	Present, elongate	90-95%

Adapted from Kurtböke et al. (2007).

AMF infections observed in the Fraser Island sand dune samples from the rehabilitated area such as (i) looping hyphae were present near the point of entry into the cortex of the fine feeder roots of *C. equisetifolia* var. *incana.*

which is the diagnostic of infection by *Gigaspora* sp., (ii) presence of lobed vesicles and numerous large extra-matrical spores were found, characteristic of the typical of infections caused by *Acaulospora* species, confirming the survival of these associations in the previously disturbed areas. Figure 5 illustrates the entry of AMF hyphal filament into the fine feeder roots and numerous vesicles in the cortical cells of the *Spinifex sericeus* found on the island. The shape of these vesicles are typical of infection by *Glomus* species reported by other researchers. (see http://www.soilhealth.com/fungi/02.htm).

In addition, Authors recovered *Penicillium* species in the study areas of the Fraser Island. There was significantly more colony forming units of fungal species in the mature hind dune community than in any other soil examined. Evidence was presented earlier that AM can work synergistically with organisms such as P solubilising bacteria and fungi to further increase available P in the rhizosphere. This has been especially highlighted in sterile soils where plants did noticeably better when AM inoculation worked synergistically with the phosphate solubilising fungus *Penicillium balaji* (Smith and Read 1997). Perhaps lower numbers of *Penicillium* species detected in the rehabilitated areas might support the findings of Smith and Read (1997).

Occurrence of actinomycetes in the natural environments might also correlate with the limited colonization rate of the *Casuarina*-plants in the rehabilitated hind dune zone.

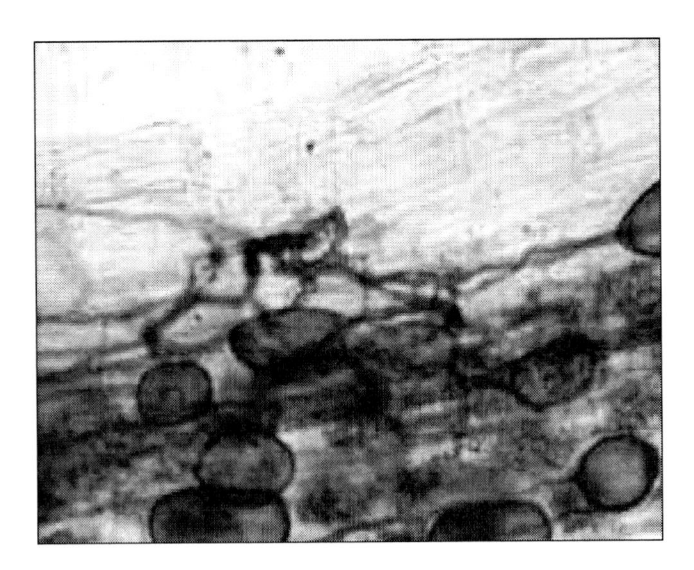

Figure 5. AMF association with *Spinifex sericeus*.

Actinomycete species (Figures 6a,b and 7a,b,c,d) in particular, micromonosporae and streptomycete species isolated by the Authors on Fraser Island (Kurtböke *et al.* 2007), belong to taxa previously reported to produce plant beneficial compounds. Failure to detect culturable actinomycetes or detecting these in negligible numbers in the rehabilitated areas using conventional techniques might result from previous disturbance to sand in the course of mining activity, when large tracks of dune were cleared and exposed for lengthy periods. As a result, the sands remained un-vegetated and subjected to similar processes as natural sand blows removing beneficial microflora from the upper layer of the sand which is the vital point of microbe-plant interaction. Apparent lack of large numbers of neutrophilic actinomycetes might be due to long-term effects of exposed sand dunes in these areas (Table 3) as well as alkali sand conditions. Large numbers of alkaliphilic/alkalitolerant streptomycetes have been isolated from sand collected from the beach and sand dune systems of Northumberland (UK) (Antony-Babu *et al.* 2008). All strains were taxonomically characterized as members of the genus *Streptomyces.* Subsequent examination of the extracts from 29 of these alkaliphilic streptomycete strains (from Northumberland) and identified a remarkable number and variety of new secondary metabolites (Fiedler and Goodfellow 2004). So, the presence or absence of these species may have implications in relation to plant growth promotion as well as being involved in the supply and/or availability of limiting nutrients in sand dunes such as nitrogen.

Furthermore, the apparent lack of actinomycete species in the rehabilitated sand dune zones may be implicated in the lower percentage of AMF root colonization (Table 2 and 3). Arbuscular mycorrhizal fungi are known to be influenced by the activity of other members of the rhizoplane/rhizosphere actinomycete populations (e.g. Ames et al. 1989). The role of actinomycetes as plant growth promoters as well as mycorrhizal facilitators has been demonstrated (Riedlinger *et al* 2006, Schrey *et al* 2007). Actinomycetes were isolated from the spores and hyphae of the AMF *Glomus macrocarpum* (Ames *et al.* 1989) and these actinomycete isolates were shown to be beneficial to the AMF development and growth of onions. The correlation between the absence of actinomycetes and the decreased rate of development of the mycorrhizal association in the roots of pioneer plants of Fraser Island might result in poorly reconstructed zones due to the absence of "mycorrhiza-helper actinomycetes. Recently, the so-called, "mycorrhiza helper bacteria" *Ralstonia basiliensis* and *Bacillus subtilis* have been identified as possessing the ability to enhance the mycorrhizal symbiosis between *Suillus granulatus* and *Pinus thunbergii* (Kataoka *et al.* 2009).

Table 2. Comparison of the numbers* of streptomycetes, other bacteria and fungi isolated from Fraser Island sand dunes

Dune Type	Streptomycetes*	Other Bacteria	Fungi
Fore dune			
	ND	4X103	ND
Intact (mature) hind dune			
	$4.2X10^3$	$5.2x10^3$	$3.8x10^3$
Rehabilitated hind dune			
	ND	$6x10^3$	$1.1x10^3$

*Colony forming units/g air dried soil.

ND: Non-detected when conventional isolation techniques and neutrophilic conditions are used.

Table 3. Types of culturable family members of different Actinomycete suborders detected in the sand dunes of the Fraser Island

Sampling Site	Sub-order	Family
Beach Zone	ND	
Foredune	Micromonosporineae	Micromonosporaceae
Successional dunes	Streptomycineae	Streptomycetaceae
Rehabilitated dunes	ND	
Mature Hind Dune	Streptomycineae	Streptomycetaceae
	Pseudonocardineae	Pseudonocardiaceae
	Corynebacterineae	Nocardiaceae
	Streptosporangineae	Thermomonosporaceae
	Propionibacterineae	Nocardioidaceae
	Frankineae	Geodermatophilaceae

ND: Non detected.

While non-native species such as fast-growing grasses were traditionally used to stabilise sand dunes, the stability came at the expense of local native species that had evolved to survive, rather than compete. Particularly, when propagated from local seeds, native species provide a sound alternative for revegetation (Greening Australia Queensland (Inc.) 2001). In spite of their importance, in depth studies have not been conducted in the microflora analysis of the sand dune plant rhizoplane and rhizosphere, so far. Benefits from light disturbances such as trampling or sand burial in stimulating the growth of stabilising plants such as spinifex were reported by Greening Australia Queensland (Inc.) (2001) and might be due to the introduction of plant growth promoter microflora.

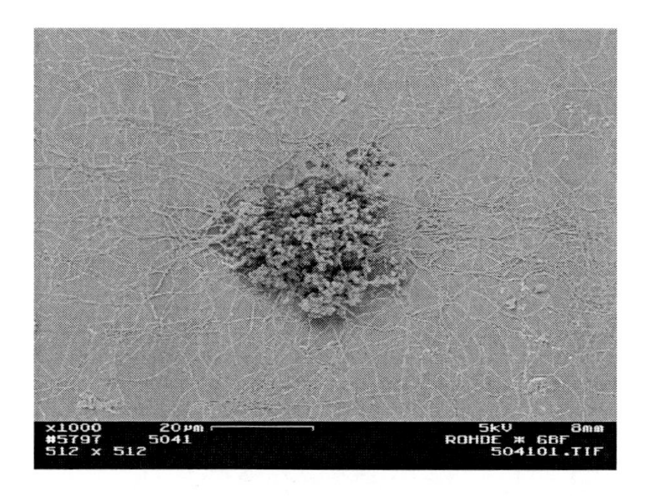

(a) View of substrate mycelium and single spore clusters of the isolate.

(b) Close view of single spores of the isolate.

Figure 6. Electron micrographs of a *Micromonospora* species isolated from sucessional dunes of Fraser Island, Queensland, Australia.

Until recently the effect of AMF on the growth of dune-colonising plants and on dune stabilisation has received almost no study. Considering the dramatic response of many crop and ornamental plants to inoculation with AMF when growing in phosphorus-deficient soils (e.g. Smith and Read 1997), the presence of AMF in post-disturbance dune vegetation would be likely to enhance plant nutrition and sand-binding (Read *et al.* 985; Logan *et al.* 1989). Greenhouse studies have shown a slight but significant increase in beach-grass *Ammophila* sp. plant growth in response to AMF inoculation (Koske and Polson 1984). However, the benefits of AMF colonisation extend beyond plant growth responses, and planting tube-stock of beach-grasses and other foredune vegetation with roots already colonised by indigenous AMF may permit the plant to grow vigorously in locations with low densities of AMF spores.

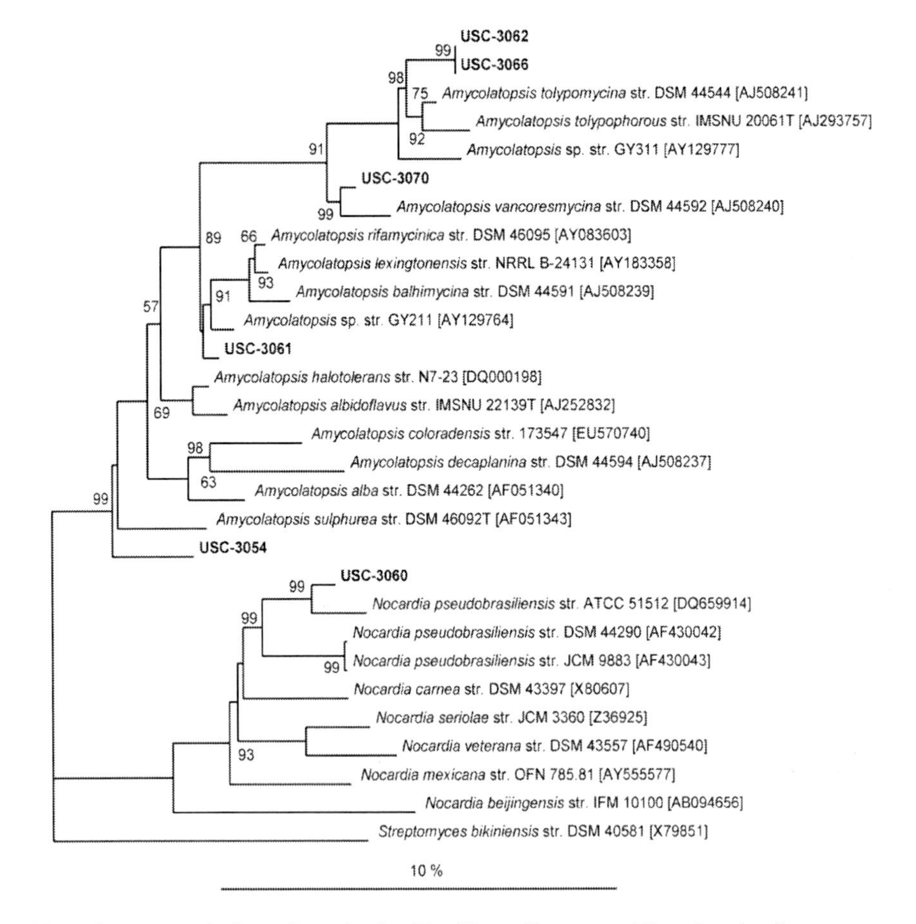

(a) Actinomycete isolates from the families Nocardiaceae and Pseudonocardiaceae

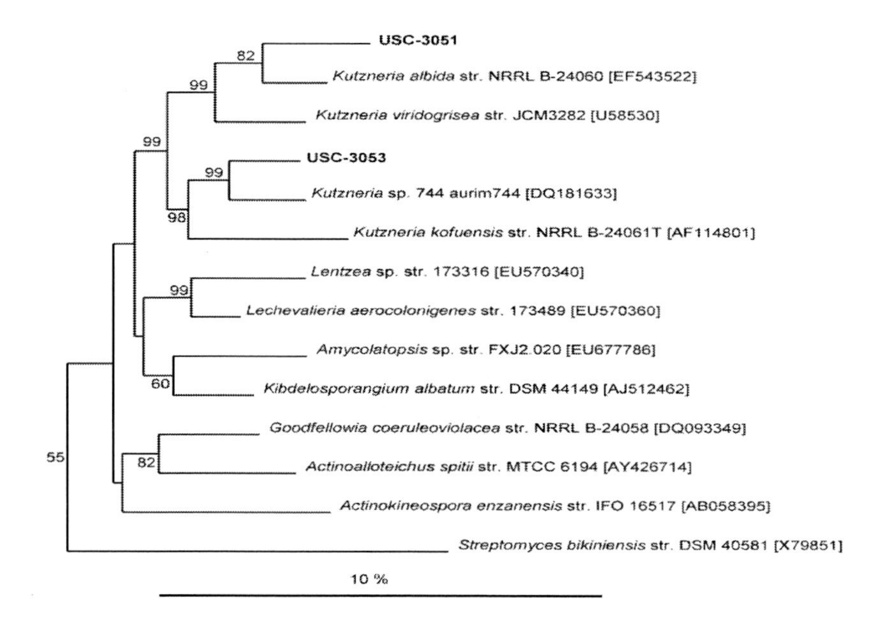

(b) Actinomycete isolates from the family Pseudonocardiaceae

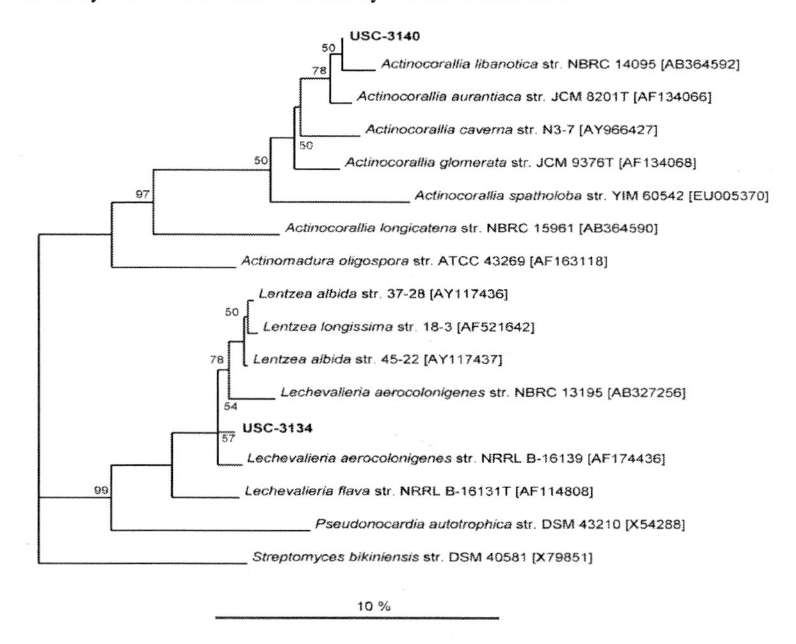

(c) Actinomycete isolates from the families Thermomonosporaceae Actinosynnemataceae

Figure 7. Continued on next page.

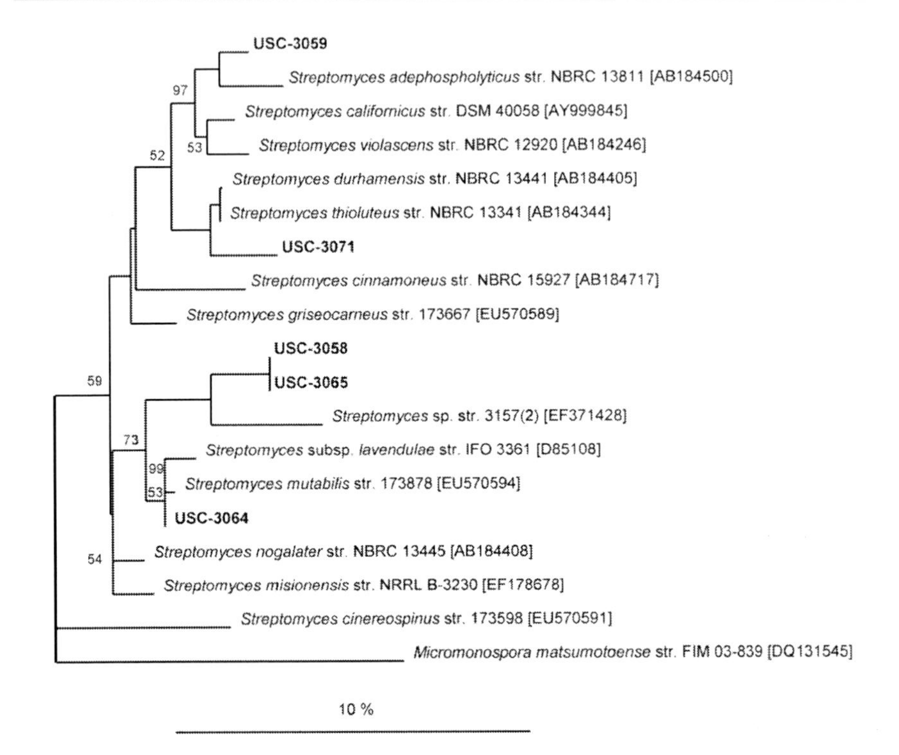

(d) Actinomycete isolates from the family Streptomycetaceae

Figure 7. Phylogenetic tree of 16s rDNA gene sequences of actinomycete isolates from mature hind dunes of Fraser Island (a-d)). The scale bars represents 10% sequence divergence. Genebank accession numbers of reference sequences are presented.

Conclusions and Future Prospects for Dune Stabilisation

The poorly developed canopy and reduced diversity in the herbaceous layer of the rehabilitated hind dune suggests that the recovery process is slow in disturbed sand dunes both in density and diversity at Fraser Island. Continuous analysis and monitoring of changes in the microfloral diversity are needed to be able to relate the speed and trajectory of the plant community development to the diversity and abundance of actinomycete and AMF populations. Direct comparisons with intact native bushland can assist with providing baseline information on the microbial complement of the undisturbed ecosystem. In this way, the threshold levels of actinomycete and

AMF inoculum could be determined and information could be obtained on what background levels of these symbionts are necessary to achieve sustainable development of long-term, stable vegetation associations after disturbance associated with sand mining.

Augmentation of poorly recovered areas with plant beneficial microflora might be considered to increase the establishment rate of the plants in revegetated areas. Commercial bio-inoculants are becoming more readily available, but understanding the species that comprise the actinomycete and AMF populations is more important, as these systems have undergone extreme selection bottle-necks to achieve the plant-microbe associations that we currently observe. Through the use of molecular-based technologies and Real-Time PCR, more information can be gained on those unculturable symbiotic microorganisms. Obligate biotrophs such as AMF fungi can be identified through the extraction of genomic DNA from fine roots or root nodules and using universal primers (e.g. White *et al.* 1990), or the new Glomalean-specific primers (e.g. Redecker 2000), to amplify the nuclear small subunit (18s) and internal transcribed spacers (ITS).

Present evidence suggests that AMF are of vital importance to the survival and growth of the major dune-colonizing plant species, and thus, the dunes themselves (Koske and Polson 1984). Mechanical, environmental and/or chemical factors that reduce or eliminate the fitness of AMF and other symbiotic micro-organisms such as actinomycetes may have serious implications for the long-term stability of rehabilitated dunes systems. Without the establishment of the deep-rooted, key-stone, hind-dune *Casuarina* trees, the ability of the dune-systems to withstand on-going coastal erosion events may be undermined. *Casuarina*-trees also provide essential ecosystem services to the wildlife that inhabit the dunal regions of this World Heritage Site, as well aesthetic values which enhance the coastal vista.

Acknowledgment

Authors acknowledge the past, present and future Traditional Custodians of the Island K'Gari "The Badtjala People" and their Ancestors. Authors also gratefully acknowledge Assoc. Prof. Ron Neller's expert advice and contribution to the research study on Fraser Island field trips and James Harper for technical help. Authors also thank Dr Manfred Rohde at the GBF, Germany for the electron micrographs of the actinomycete species and Dr Ken

Wasmund for the construction of phylogenetic trees. SEB's research tenureship was supported by 2004/2005 USC-research grant UR604/4.

REFERENCES

Ames RN, Mihara KL and Bayne HG (1989). Chitin decomposing actinomycetes associated with VAM fungus from a calcareous soil. *New Phytologist,* 111: 67-71.

Anthony-Babu S, Stach JE and Goodfellow M. (2008). Genetic and phenotypic evidence for *Streptomyces griseus* ecovars isolated from a beach and dune sand system. *Antonie van Leeuwenhoek*, 94(1): 63-74.

Bâ AM, Diédhiou AG, Prin Y, Galian A and Duponnois R (2010). Management of ectomycorrhizal symbionts associated to useful exotic tree species to improve reforestation performances in tropical Africa. *Annals of Forest Science*, 67(3): 301-310.

Becking (1977). Dinitrogen-fixing associations in higher plants other than legumes. In: R. W. F. Hardy and W. Silver, (eds), Section III: Biology, *A Treatise on Dinitrogen Fixation*, pp. 185-276. NY: John Wiley and Sons, NY.

Bellgard (1992). The propagules of VAM fungi capable of initiating VAM infection after topsoil disturbance. *Mycorrhiza,* 1: 147-152.

Bellgard SE (1993) Soil disturbance and infection of *Trifolium repens* roots by VAM fungi. *Mycorrhiza*, 3: 25-29.

Brockoff JO and Allaway WG (1989). VAM fungi in natural vegetation and sand-mined dunes at Bridge Hill, NSW. *Wetlands,* 8(2): 47-54.

DEHWA (2010). Fraser Island Information Sheet (accessible at: http://www.environment.gov.au/heritage/education/pubs/factsheets/fraser-island.pdf).

Dell, B (2002). Role of mycorrhizal fungi in ecosystems. *CMU Journal,* 1(1):47-60.

El-Tarabily KA, Sykes ML, Kurtböke DI, Hardy GEStJ, Barbosa AM, Dekker RFH (1996). Synergistic effects of a cellulase-producing *Micromonospora carbonacea* and an antibiotic producing *Streptomyces violascens* on the suppression of *Phytophthora cinnamomi* root rot of *Banksia grandis. Canadian Journal of Botany*, 74: 618-624.

Dobson AP, Bradshaw AD and Baker AJM (1997). Hope for the future: Restoration Ecology and Conservation Biology. *Science,* 277: 515-522.

Fiedler H-P and Goodfellow, M (2004). Alkaliphilic actinomycetes as a source of novel secondary metabolites. *Microbiology Australia*, 25(2): 27-29.

Gardner JH and Malajczuk N (1988). Recolonization of rehabilitated bauxite mine sites in WA by mycorrhizal fungi. *Forest Ecology and Management,* 24: 27-42.

Greening Australia Queensland (Inc.) (2001). Coastal Dune Vegetation. Fact Sheet Series Published with the assistance of the *Bushcare*, A program of the Commonwealth Government's Natural Heritage Trust. Produced by GAQ Training Development Division.

Hill TCJ, Tippett JT, Shearer BL (1994) Invasion of Bassendean Dune *Banksia* Woodland by *Phytophthora cinnamomi. Australian Journal of Botany,* 42(6): 725–738.

Jasper DA, Abbott L and Robson AD (1991). The effects of VAM fungi in soils from different vegetation types. *New Phytologist,* 118: 471-476.

Jasper DA, Abbott LK, Robson AD (1992) Soil disturbance in native ecosystems, the decline and recovery of infectivity of VA mycorrhizal fungi. In: DJ Read, DH Lewis, AH Fitter, and AJ Alexander (eds) *Mycorrhizas in Ecosystems*, CAB International, Wallingford, UK, pp. 151-155.

Jehne W, Thompson CH (1981). Endomycorrhizae in plant colonization on coastal sand-dunes at Cooloola, Queensland. *Australian Journal of Ecology*, 6: 221-230.

Klironomos JN and Moutoglis P (2004). Colonization of nonmycorrhizal plants by mycorrhizal neighbours as influenced by the collembolan, *Folsomia candida. Biology and Fertility of Soils*, 29(3): 277-281.

Kataoka K, Taniguchi T and Futai K (2009). Fungal selectivity of two mycorrhiza helper bacteria on five mycorrhizal fungi associated with *Pinus thunbergii. World Journal of Microbiology Biotechnology*, 256: 1815-1819.

Koske RE and Polson WR (1984). Are VA mycorrhizae required for sand dune stabilisation? *BioScience*, 34(7): 420-424.

Kurtböke DI, Neller RJ and Bellgard SE (2007). Mesophilic actinomycetes in the Natural and Reconstructed Sand Dune Vegetation Zones of Fraser Island, Australia. *Microbial Ecology*, 54(2): 332-340.

Lemauviel S and Roze F (2003). Response to three plant communities to trampling in a sand dune system in Brittany (France). *Environmental Management*, 31(2): 227-235.

Logan, VS, Clark PJ and Allaway WG (1989). Mycorrhizas and root attributes of plants of coastal sand-dunes of New South Wales. *Australian Journal of Plant Physiology*, 16: 141-146.

Lubke RA (2008). Vegetation dynamics and sucession on sand dunes of the Eastern Coasts on Africa. In: MI Martinez and MP Psuty (eds) *Coastal Dunes, Ecology and Conservation. Ecological studies*, 171(II): 67-84. Springer: Verlag, Berlin, Heidelberg, 2004.

Maremmani A, Bedini S, Matosevic I, Tomei PE and Giovannetti M (2003). Type of mycorrhial associations in two coastal nature reserves of the Mediterranean basin. *Mycorrhiza*, 13(1): 33-40.

Park MS, Jung SR, Lee MS, Kim KO, Do JO, Lee KH, Kim SB and Bae KS (2005) Isolation and characterization of bacteria associated with two sand dune plant species, *Calystegia soldanella* and *Elymus mollis*. *Journal of Microbiology*, 43(3): 219-227.

Peterson RL, Ashford AE and Allaway WG (1985). VAM associations of vascular plants on Heron Island, a Great Barrier Reef Coral Cay. *Australian Journal of Botany*, 33: 669-676.

Ponder F Jr. (1980). Rabbits and grasshoppers: vectors of endomycorrhizal fungi on new coal mine spoil. *Forest Service Resource Note* NC-250, USDA Forest Service, Washington.

Read DJ, Francis R and Finlay RD (1985). Mycorrhizal mycelia and nutrient cycling in plant communities. In: AH Fitter, D Atkinson, DJ Read and MB Usher (eds) *Ecological Interactions in Soil – Plants, Microbes and Animals*, pp 193-217. Blackwell Scientific Publications,

Redecker D (2000). Specific PCR primers to identify arbuscular mycorrhizal fungi (Glomales) within colonized roots. *Mycorrhiza*, 10: 73-80.

Riedlinger J, Schrey SD, Tarkka MT, Happ R, Kapur M and Fiedler HP (2006) Auxofuran, a novel metabolite that stimulates the growth of fly agaric, is produced by the mycorrhiza helper bacterium *Streptomyces* strain AcH 505. *Applied and Environmental Microbiology*, 72(5): 3550-3557.

Rodriguez-Echeverria, S and Freitas, H. (2006). Diversity of AMF associated with *Ammophila arenaria* ssp. *arundinacea* in Portuguese sand dunes. *Mycorrhiza*, 16(8): 543-552.

Schrey SD, Schellhammer M, Ecke M, Hampp R and Tarkka MT (2005). Mycorrhiza helper bacterium *Streptomyces* AcH 505 induces differential gene expression in the ectomycorrhizal fungus *Amanita muscaria. New Phytologist,* 168: 205-216.

Shearer BL and Dillon M (1996). Impact and Disease Centre Characteristics of *Phytophthora cinnamomi* Infestations of *Banksia* Woodlands on the Swan Coastal Plain, Western Australia. *Australian Journal of Botany,* 44(1): 79–90.

Shin DS, Park MS, Jung S, Lee MS, Lee KH, Bae KS and Kim SB (2007). Plant growth promoting potential of endophytic bacteria isolated from roots of coastal sand dune plants. *Journal of Microbiology and Biotechnology*, 17(8): 1361-1368.

Smith, SE and Read DJ (1997). *Mychorrizal Symbiosis,* 2nd ed. San Diego, California: Academic Press Inc.

Solans M (2007). Discaria trinervis–Frankia symbiosis promotion by saprophytic actinomycetes. *Journal of Basic Microbiology*, 47: 243-250.

Sutton JC and Shepherd BR (1976). Aggregation of sand dune soil by endomycorrhizal fungi. *Canadian Journal of Botany,* 54: 326-333.

Tan RX and Zou WX (2001) Endophytes: A rich source of functional metabolites, *Natural Products Reports*, 18: 448-459.

Tisdall JM and Oades JM (1979). Stabilization of soil aggregates by the root systems of ryegrass. *Australian Journal of Soil Research*, 17: 429-441.

White TJ, Bruns TD, Lee S and Taylor JW (1990). Amplification and direct sequencing of fungal ribosomal RNA genes for phylogenetics. In: MA Innis, DH Gelfand, JJ Snisky and TJ White (eds) *PCR protocols: A Guide to Methods and Applications,* pp. 315–322. Academic press, San Diego.

In: Sand Dunes ISBN 978-1-61324-108-0
Editor: Jessica A. Murphy, pp. 119-135© 2011 Nova Science Publishers, Inc.

Chapter 5

THE SAND DUNE SYSTEMS IN THE NORTHERN COAST OF SENEGAL: ORIGIN, CHARACTERISTICS, ECONOMIC AND ENVIRONMENTAL IMPACTS

Mamadou Fall** *and Mouhamadoul Bachir Diouf
Département de Géologie, Faculté des Sciences et Techniques,
Université C.A.Diop de Dakar, Sénégal

ABSTRACT

The Quaternary sedimentary basin of Senegal is marked by phases of edification of varied dune systems, differently distributed in time and space. Three systems are identified. Akcharien erg, prior to 100 ka BP, leveled or more or less dismantled in places. Ogolian erg built along the Atlantic coast of Senegal and Mauritania during the arid phases of Würm III. Barrier beaches over Holocene and Present, built during the regularization of shorelines. These dune systems mainly contain quartz and useful heavy minerals relatively wellpreserved in the barrier beaches. Climate change of the region and eustatic fluctuations have led to a particular pattern of Ogolian sand dunes conducive to the installation of a groundwater aquifer and lush vegetation in areas now transformed into interdunes market-gardening. The drought of the 70s and anthropogenic activities have led to degradation of ecosystems including the marked

---

* mfall313@yahoo.fr

decline of groundwater levels, deforestation, remobilization of dune fronts and sanding up farming areas.-

INTRODUCTION

The establishment and evolution of the constituent materials of dunes are a component of the rock cycle. The range of ferromagnesian or silicate minerals constituting a dune deposit, by their mineralogical nature and degree of wearing are the marks of major geodynamic events, some internal (magmatism, metamorphism, tectonic deformation) and other externally (erosion, wind and water transport, sedimentation, diagenesis). What particularizes dunes is their direct exposure to the action of natural environmental factors (climatic factors) and anthropogenic factors. Most decisive natural factors are surface water (rain or sea water, sea spray) and winds.

However, the consequences of the action of winds, which are the main motor of the mobility of dunes and their modeling, can be reduced or canceled out by the second factor, water, especially when it leads to the appearance of colonial vegetation that inhibit the mobility of dunes, stabilize slopes and reduce the gradients. In short, natural evolution of the dune deposits appears as a competition between the agents responsible for the erosion dynamics and mobility modeling of dunes and pedo-geochemical processes induced by the presence of water and which may lead to install an environment favorable to the conservation of dune building.

Our study area, the western region of Senegal , is located on the Atlantic seaboard opened to trade maritime winds and continental trans-Saharan trade winds and is subjected to a regime of seasonal rainfall of intensity highly variable over the last millennia (Fig. 1). This is an area covered by large clumps of dunes of various ages whose study was based on sedimentological, paleontological, botanical and isotopic markers. Dune sets, most azoic, were included in the International Stratigraphic chronology. However, very often they local stages are used: Ackarian (1.000 000 BP), Ogolian (30.000 to 20.000 BP), Tchadian (10.000 to 6.800 BP) and Tafolian (4.200 to 2.000 BP).

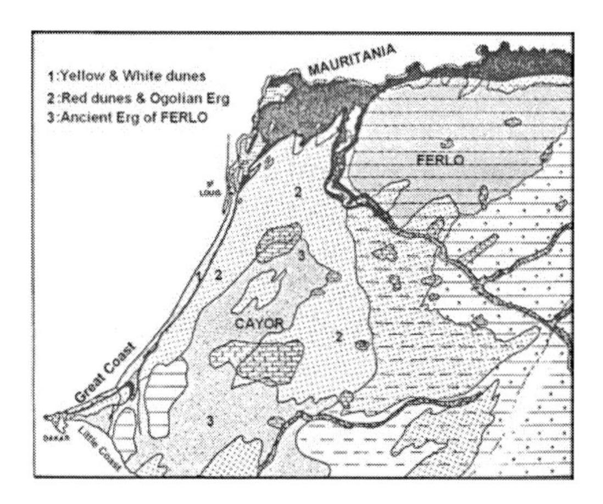

Figure 1. Quaternary in SENEGAL (in ATLAS NATIONAL 1977)

The rapid climatic changes that are occurring under our eyes in recent decades provide an opportunity to observe the geological features of these dunes and processes of morphological and pedological evolution. Thus it was possible to follow both the recent dune-building process, and various states of evolution of old dunes, including phases of mobilization and fixation, changes in association with availability of stormwater and stages of pedological colonization that led to the formation of specific ecosystem called "Niayes" (Fig.3). These richly forested wetland ecosystems as well as the Atlantic coast nearby certainly played a role in the installation, for millennia, of human settlements in the vicinity of these sites.

Human activities then come in play, sometimes to counter (timidly) erosion factors (reforestation and dune fixation), but more often to increase the degradation of the environment. The installation of cities, agriculture, grazing and extraction of valuable materials (sands, heavy minerals) involved rather to increase desertification of sites sorely tried by lower rainfall observed since the early 70s.

2. Origin and Morphosedimentary Characteristics of Dunes of Senegal West

Glacio-eustatic and climatic Quaternary episodes were deeply marked in the geology of the western margin of the basin and have been identified through numerous studies [1- 6].

Climatic changes had a major role in the morphogenesis of the Senegal-Mauritanian Basin. Climatic and eustatic variations have printed the region a singular evolution. Marine transgressions generally coincide with wet long periods while the regressions occurred during dry phases often shorter. The region has known a climate evolution marked by a clear correlation between high sea levels and storm sewers that astronomical theory of Milankovitch allows liaise with the glacial and interglacial phases of the northern hemisphere [7-8].

Glacial periods corresponded, in the western part of Africa with aridity phases more or less strongly marked by a significant detrital sedimentation source of the various dune formations.

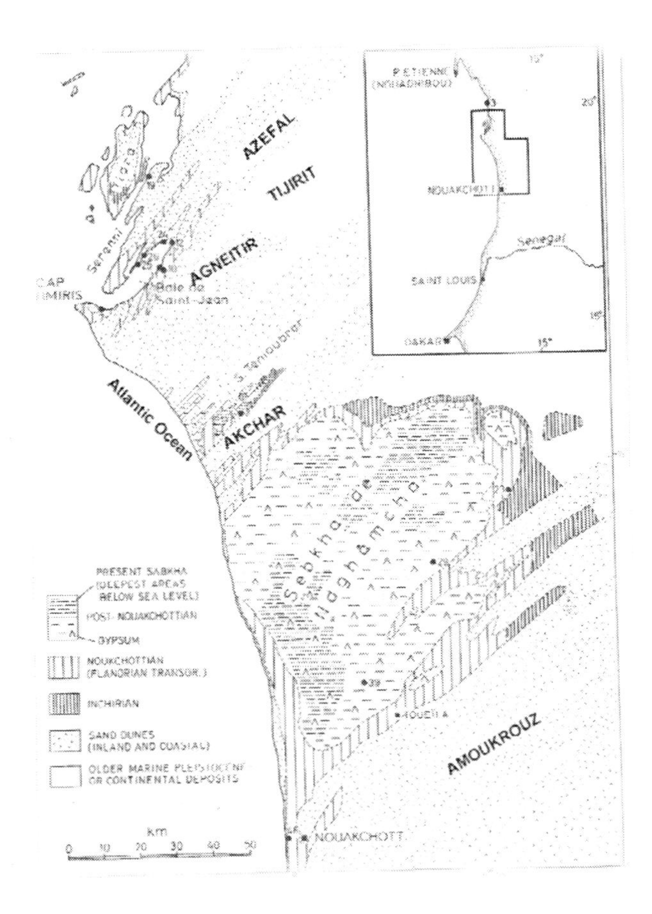

Figure 2. Geology of Westhern Mauritania

2.1. Ancient Erg

It straddles Mauritania, Ferlo and Cayor in central Senegal (Fig.l). Are grouped under this term transverse dune formations, North-South direction, built during the maximum of aridity following the building of medium glacis in the basin of the Senegal River. This long dry spell has accompanied the post Tafaritian regression (Akcharian). It appears that there were indeed several phases of aridity all associated with phases of cooling and highlighted by changes in *Melosira*. Continental diatoms (lakes, rivers) of the genus *Melosira* are an indicator of aridity in so far as their development reflects a general deflation and a strong drying trend lakes. Studies indicate that high contents in biological origin aerosols are located in glacial settings supported by the curve of oceanic oxygen-18 isotope. Two of *Melosira* peaks highlighted here correspond to Stage 5 *i.e* to the Eemian interglacial period (around 125.000BP) [11].

Erg formation which characterizes this Akcharian episode therefore comes during an arid stage correlated to the glacial period corresponding to isotope stage 5 and appears to have been promoted by Tassiast and Tijirit regions uplift in Mauritania [5] (Fig. 2) and subsequent strengthening of erosion.Akcharian sand dunes have low slopes and a dull increasingly marked towards the south. They are composed of sands with river marked characters, mode 0.8mm [8] , heavy minerals abundant and variously distributed along a NE-SW. [9], describe clear field distributions of heavy mineral and find a good correlation of variations found with the course of ancient rivers that connected the inland regions to coastal areas. The nature of minerals (amphibole, epidote, staurolite, kyanite, etc..) characterize the sources of the material (Amsaga region, Mauritanides).

Conditions of ancient erg implementation seem quite so easy to establish. Erg is composed of accumulations of material shortly transported by winds manifestly not strong.

2.2. The Ogolian Sand Dunes

Unlike old dunes erg, winds of NNE -SSW direction causing Ogolian forms are more violent [10]. Thus they conveyed over long distances large quantities of sediment which witnesses are visible both on the Great North Coast of Senegal than on the Little South Coast. These are longitudinal dunes NE-SW directed ranging over twenty kilometers long and 35-40 meters high.

They cover a large part of Senegal and consist of a sandy sediment (0.16 - 0.25 mm), reddened, with very low clay fraction. They are put in place during the climatic aridification that accompanied the last great period of regression in Senegal and Mauritania basin (20 000 - 15 000 BP). This implementation was done under arid lowstand.

Removal could reach a level close to -120 m. It was at this time that formed in Mauritania to dune ridges of Agnéïtir, Amoukrouz, Akchar and Azefal [5]. In fact, this would be the last Ogolian erg representative of a series of "ice desert" [11] developed between 70 000 and 12 000 BP), in an environment less warm that during the interglacial periods. The quartzy sediment exoscopy shows traces of redepositions [8] (at least 2) suggesting multiple phases of remobilization. Redepositions phases would have occurred between 7000 and 6500 BP [7] , due to a driest episode, causing very local remodeling. This remobilization of Ogolian sand dunes is also the source of smaller coastal dunes. during the Tchadian (10 000-7 000 BP) Ogolian dunes underwent pedological changes leading to reddening of sedimentary material. The Thadian episode, more humid, will also stabilize sediments by dune vegetation.

2.3. Barrier Beaches

Senegalese coastal geomorphology, particularly on the great northern coast, has three dune systems. The coastal dunes also called a) white dunes: active dunes frequently remobilized with bioclastic sandy material, subactual to recent (2000 - 1800 BP). They are locally colonized by halophytic vegetation. b) Semi-fixed yellow dunes in the background of active dunes. They are locally interrupted by lakes (Retba, Mbeubeuss, Youi, Malika etc). The red sand dunes inland or inner dunes belonging to the Ogolian erg previously described above (Fig. 3).

These one are covered by a large vegetal cover forming by places savanna woodlands. They consist of juxtaposition of two morphological units different by altitude of sand dunes summits, importance of the system crests - dune valley and by dunes orientation [12]. Inward Ogolian dunes are long and leveled. However, the original alignment of NE direction, parallel to the coast is still observable despite clogging dune valley. Away, Ogolian dunes were reshaped during Tchadian (Holocene) wet phase. They are short and have a particularly complex form characterized by relatively high elevations in some areas (more than 30 m in some areas), deep dune valley with digitate contours, often tight and crosslinked around parabolic dunes oriented in directions ranging from NW and NNE directions to straight forward East.

Yellow semi-fixed dunes and white active dunes constitute outer system of current coast. Field yellow dunes have been implemented during post-nouackchottian regressive phases (after 5500 BP). They are built to Tafolien (4.2 Kyr BP) in favor of powerful longshore currents that regulate the coast and aeolian sedimentation influenced by strong winds. A drying climate occurred around 4Kyr followed by a wet phase around 3.5 Kyr as shown by palynilogic analysis of fossil peats [13]. Furthermore, paleoclimatic changes in the northern coast of Senegal during the Tafolian are investigated with ^{13}C method, on a peat core of 350 cm long spanning the past 3,500 yrs extracted from an interdune basin located at 15.07° N, 16.53°W [14]. This deposit corresponds with a local stage, the Tafolian (4,000 BP to 2,000 BP) whose paleoclimatic trends, essentially defined by sedimentological methods, were previously considered to be dominated by drought. The ^{13}C profile obtained from 70 successive peat samples shows climate oscillations between wet and dry periods. A wet period appears around 3,200 BP. A shorter but equally wet one occurs around 1,000 BP. These two humid phases are separated by a rather arid period that occurs around 2,000 BP as shown by an exceptional rise in ^{13}C values. This dry phase as well as a second one evidenced around 800 BP coincided with increases of dull-quartz abundance; originating from wind-reworking of surrounding dune sand. After 850 BP, the carbon isotope record suggests a progressive return to more humid conditions. Carbon isotope trends have shed new light on past monsoon rainfall oscillations over Sahelian areas during the Tafolian

This short arid period allowed accumulation by maritime trade winds of these dunes with poorly developed soil which profile reduces to the humus horizon. The width of the field dunes reaches 4 km in some areas, while it narrows considerably downstream of major depression-like stream. The most massive dune forms have been ten meters high and about ten kilometers in length. Their general orientation is NE-SW. They form in the most northern coastal areas, in particular in the old mouths of Senegal River, parabolic dunes [3] further south in the Cap Vert region are installed dunes with complex shapes mainly represented by the Camberene erg. Leeward slopes of dunes overlook the erg Ogolian by steep slopes which become very scored at the base of the depressions. A clear trend remobilization is felt now and in some places reactivated dunes tend to bury depressions (Fig.3).

Dunes are composed of uni-modal sedimentary material (mode 0.11 mm; [8], shelly and rich in heavy minerals [5]. Such as all other wetter phases, it enabled the functioning of numerous lakes that occupied the dune valley areas and which were source of significant lake deposits including peat. Recently active dunes formed near the shoreline a belt of a few meters to several hundred meters with a relatively complex morphology. At the back beach, under winds action, beach ridge is organized into small parabolic dunes. Altitudinal contrasts are so hard towards this level. In the opposite, inward, connection with the yellow dunes is done generally gently sloping.

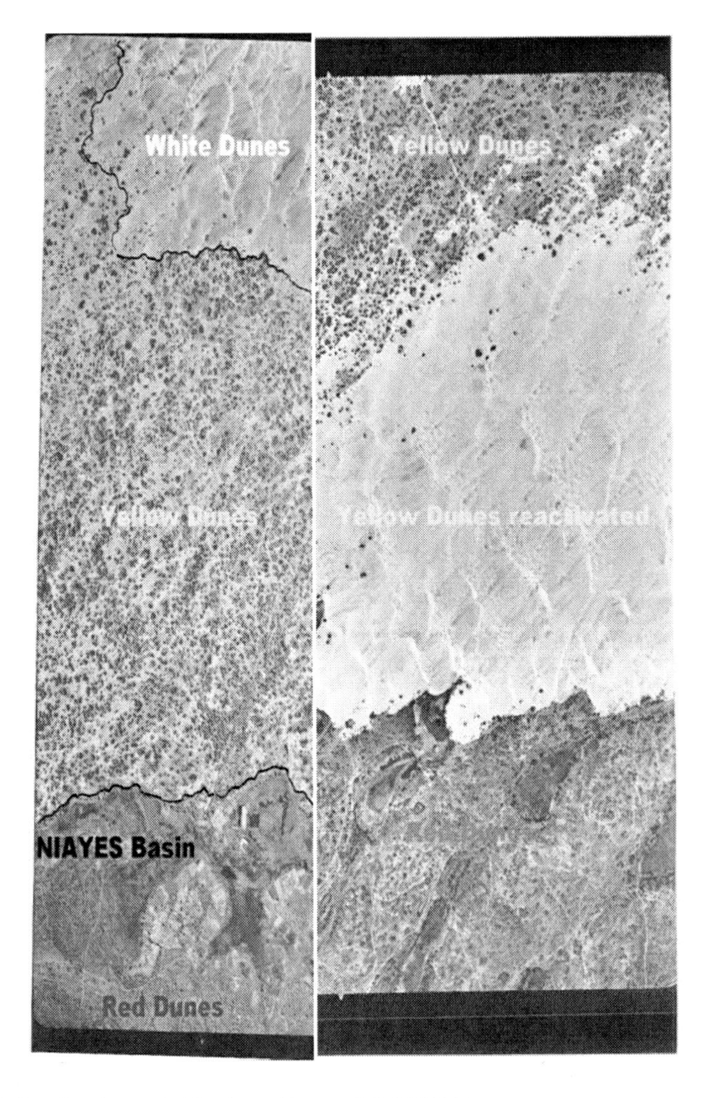

Figure 3. Barrier dunes in Northern Coast of SENEGAL

3. Quaternary and Recent Environmental Changes of Senegalese North Atlantic Coastal Dunes

As mentioned in the introduction, what particularizes dunes is their direct exposure to the action of natural environmental factors and anthropogenic factors that act as soon as the sediments are exposed on the continental environment. In our study area, the pattern of stages in the evolution of the dunes after their implementation can be investigated from geomorphological, geological, botanical, palynological, pedological and hydrogeological studies conducted on these formations

3. 1. Implementation of Hydrological and Hydrogeological Conditions

Dune systems described above provide an important freshwater aquifer based on Eocene marl and limestone. Bedrock is affected by sloping towards the north west, Thickness of sand reaches a maximum depth upright the shoreline . The origin of this water table is related to humid climate return noted at the beginning of the Holocene (Tchadian) and which followed the strong aridity that has characterized Ogolian. Improvement in rainfall has caused a gradual rise of the groundwater concomitantly to shoreline rise which passed between 20,000 BP and 7000 BP from nearly -100 m to nearly 20 m relative at its current level. Fluctuating episodes of recharge and discharge groundwater tables are highlighted by sedimentary, palynological [16] and isotopic markers [15].

Actually this coastal aquifer is exploited near 100 000 $m^3.day^{-1}$ for urban and rural populations located between Dakar and St. Louis. Piezometric data reveal two main drainage axes, the one goes towards the ocean, contributing to limit freshwater pollution by marine waters. The other led towards the marl and limestone of the middle Eocene and involved in recharging the water table water contained in these limestones. Water upraising due to stormwater inputs produce numerous lacustrine accumulations due to the exposure of free water bodies in depressions. The feeding depressions at the bottom of slopes can be made directly from the water contained in the field dunes. This process can continue long after drying of the depression center and help maintain an almost permanent body of water in the edges of some large basins in which vegetation grows vigorously.

3.2. Vegetation Settlement

The botanic current framework is probably a very advanced stage of degradation of vegetal formations which are gradually installed on the dunes since the return of rainy episodes at the beginning of the Holocene (Fig;). However, through the botanical and palynological studies [16] [17], one can perceive what were their characteristics until the early 60s. The study of nature and evolution of these plant formations within recent interdune depressions being colonized by plants is very interesting to understand the process of installation of rich and varied plant communities " that generated the peat deposits within " Niayes" [13] [16].

The installation and evolution of plant colonies on the substrates depends on the height, morphology and mobility of dunes and the availability of fresh or salt waters. Vegetation settled on fixed or semi-fixed dunes are mainly composed of grasses and psammophilous sedges adapted to substrates more or less mobile and with few nutrients

Thus, from the coast inland, we define several successive groups [17]: a halophilic *Cyperus conglomeratus* group on the white recent dunes more or less mobile, a steppic *Aristida hyperrhenia* group on the yellow dunes more or less fixed, on the red dunes a *Aristida* group enriched by continental big tree species such as *Toningii Ficus, Ficus congensis, Detarium senegalensis, Voacanga africana.*

The edges of yellow dunes overhanging some major depressions still sustain a bushy shrub *with Fagara xanthoxyloides, Alchornea cordifolia.*The edge of depression sustained a relatively dense vegetation that remains forestry gallery permanently inundated by water from the cloth of the massive dunes that overlook depression. This groups rich plant species with Guinean affinity *Elaeis guineensis* accompanied by climbing ferns that are encountered today in tropical rainforests (0° E - 12° N) as known in Guinea. Within these formations called "Niayes" we observed in plant communities that are distributed in the space linked to the water, and over time, this appears as a progressive series what fits the deepening of basins and the slope stability [17].

On the dunes, had developed a Sahelian savanna characterized by herbaceous grasses interspersed with thorny trees (*Acacia albida, Acacia senegal, Balanites aegyptiaca, Zizyphus mauritiana)* and shrubs (*Boscia senegalensis Guiera senegalensis, Commiphora africana, Grewia bicolo*r)

3.3. Pedogenesis

Table 1. Characteristic of soil types on dunes and within « niayes » basins in the northern coast of Senegal

Soil Characters	W.D	YD	RD	NB	NC
pH	6.6	7.4	5.8	4.5	4.3
Sands 50 μ - 200μ	60.93%	49.60%	37.25%		
Silts 2 μ - 20 μ	0.0%	0.25%	1.25%	4.75	3.75
Argile	0.0%	0.0%	0.0%	4.4%	7.25
Total Carbone	0.60 ‰	0 ?38‰	1.90‰	39.70 ‰	41.62‰
C/N	3	0.9	1.7	8.1	7.68

Pedogenesis and plant colonization go hand in hand. The surface layer of sand at the interface between deep layers and the atmosphere evolved in more or less fertile soil depending on the morphology and mobility of dunes and the availability of fresh or salt waters. The structure and fertility of these soils depend mainly on the texture of the original dune substrate and magnitude of interactions between the mineral and biological environment which develops gradually. Thus, starting from the shore line to the edge and center of marshes, there are a variety of soil types with varying degrees of fertility [18] (Table 1). On the white dunes (WD), the yellow dunes (YD) and the red dunes RD), it develops raw mineral soils with sandy texture and very low levels of organique matter, while, within the borders of "niayes" (NB) and the center of "Niayes" (NC) develop moist or very wet soil relatively more fertile

4. Degradation of Dune Environments

The evolution of the flora of the dunes due to drought and human action was certainly very rapid over the last 40 years. However we have documents that date from the years 50 and 60 which can assess the degree of degradation of dune environments.

4.1. Deforestation of Marshes

The present appearance of "niayes" is characterized by a severe degradation of natural plant communities due to the combined effects of falling water table and increased cropping activities (Fig. 3). In "Niayes", the deepening the water table dries progressively more or less permanent lakes occupying the center of the depression .. Consequently, farming activities have expanded gradually from the edge towards the center of basin.

In the past, flooded "niaye" was characterized by a dominance of ferns (*Lygodium, Ampelopteris, Cyclosorus, Microlepia*). In addition to *Elaeis*, the plant community includes a few big trees such as *Ficus congensis, Bridelia micrantha*. The humid "niaye" is populated by a more rich and varied vegetation with big treesr (*Alchornea cordifolia, Aphania senegalensis Ceasalpinia Bonduca, Ficus capensis, F. ovata, F. scott-elliottii, F. vogelii, More mesozygia, Trema guineensis* ...), and climbing species (*Abrus precatorius, Adenia lobata Adenopus breviflorus, Ampelocinus multistriata, Paullinia pinnata, Rhynchosia pycnostachya ...).*

The current status shows that big woody species, ferns and climbing species have disappeared in the sites. Now, in more or less humid or swampy sites Guinean species are replaced by *Typha australis, Phragmites, Cyperus dives* surrounded by a few feet of *Elaeis*. Indeed, today, vigorous typhaies associated with reeds, sedges and grasses, can be found at most sites still not occupied by vegetable crops.

4.2. Deforestation of Dunes

The degradation of the Sahel savanna was more spectacular because of the combined impacts of reduced rainfall, extensive agriculture, pastures and bush fires. However, studies show that it is the decline in rainfall since the early 70's that is the key factor responsible for mortality in Sahelian species. Added to the patterned dune substrates has a definite impact on the evolution of landscapes. Indeed, the density of plant species is dependent of the topography of the environment. Five types of sites are identified: the top of dunes, dune slope, flat dune, bottom of dune and depression. It appears that the rate of mortality due to drought varies according to species and that for most species, mortality is declining rapidly from the top of the dune at the center of the depression, as shown by the study of the impact of the severe drought of 1972 on the steppic Savannah settled on the erg in the northern region of Senegal [19] (Table 2)

Table 2. Mortality in Sahelian species (%) depending on the topography of ancient dunes of the Erg of Senegal

Spècies	Top of dune	Slope of dune	Flat dune	Bottom Of dune	Dépression
Acacia senegal	57.8 %	53.9 %	44.4 %	52.2 %	58.6 %
Commiphora africana	28.4 %	22.5 %	6.9 %	1.7 %	1.4 %

4.3. Reactivation of the Dunes by Winds

In our study area, the remobilization of the dunes by the wind is frequent (Photo 1B. This phenomenon contributes to the degradation of soil and burial of crop areas arranged in the depressions. It is obvious that the decrease in rainfall, extensive agriculture, grazing, urbanization, together increase this phenomenon. However, the discovery of peat deposits in current and out of “niayes” basins, under sand layers of several meters thick, shows that many episodes of burial of interdune depressions were able to succeed in time

CONCLUSIONS

Actual and Quaternary dunes are detrital materials , little or no evolved, accumulated on the surface of continents. These materials have not yet been subject to geological and geochemical transformations that are the root back to the state of sedimentary rocks. The study of these materials provides an opportunity to observe and analyze their evolution in a time scale relatively short (ranging from daily scale to millennium). In the context of our study area - a intertropical Sahelian area in the Atlantic coast - the impacts of climate change are often very rapid and extensive. In sum, the dune is a geological substrate that has a high reactivity towards meteorological agents and human activities.-

References

[1] Elouard P. (1967), *Bull.Inst. Fond.Afr. Noire, Dakar, Ser. A*, t. 29, 2 p 622-836

[2] Michel P.(1969), *VIIIe INQUA Congres Vol. 1* p. 49-61 Ed. CNRS, Paris

[3] Michel P..et al. , (1968), *Bull.Inst. Fond.Afr. Noire, Dakar, Ser. A*, t. 30, p. 1-38

[4] Elouard P.et al. (1977), *Ass. Sénég. Et; Quatern. Afr. Bull. Liaison N°* 50 p. 29-49

[5] Hébrard H. (1978), *Doc. Lab. Geol Fac. Sci. Lyon N° 70*, 210p 59fig.

[6] Fall M. et al (1988), *CR Acad. Sci. Paris T. 307bSéerie II* 1773-1778

[7] Michel,P. (1973), *Mém. ORSTOM N° 63, 3 tomes* 752p.

[8] Barbey C.(1989), *Bull. Soc. Geol.* Fr. 8 (1), 13-22

[9] Swezey C. (2001) *Palaeo 167, 119-155*

[10] Diester – Haass L.(1979) *Mateor Forsch Erggebnisse Berlin* N° 31, 53-58

[11] Rognon P. (1989) *Bull. Soc. Geol. Fr.* 8 (1), 13-22

[12] Sall M. et al, (1978) *Ann. Fac. Lettre, Sci. Hum. Dakar* N°8, 197-217

[13] Fall M. and, Nongonierma A (1997) *Bull.Inst. Fond.Afr. Noire, Dakar, Ser. A,* t. 49, p. 5-15

[14] Fall M. et al, (2010) *Global and Planatary Change* 72, 331- 333

[15] Lézine A.M. (1987) *Thèse Marseille* 2 vol. 180 PP.

[16] Raynal A. (1963) *Ann. Fac. Sci. Dakar* T. 9 N°2, 121-231

[17] Ndiaye R. (2004) *Mem. Diplo. Etud. Approf. Dakar* 61pp

[18] Poupon H. (1973) *Nouvelle Editions Africaines Dakar* 96-101

In: Sand Dunes ISBN 978-1-61324-108-0
Editor: Jessica A. Murphy, pp. 137-149© 2011 Nova Science Publishers, Inc.

Chapter 6

CAUSES, IMPACTS, EXTENT, AND CONTROL OF DESERTIFICATION

Habes A. Ghrefat*
King Saud University, Department of Geology and Geophysics
Riyadh 11451, Saudi Arabia

ABSTRACT

Desertification is a serious environmental problem and it potentially affects 35% of the land surface of earth and 32% of the human population. Desertification is land degradation in arid, semi-arid and dry sub-humid areas and includes degradation of vegetation cover, soil degradation, and nutrient depletion. Overcultivation, increased fire frequency, overdrafting of groundwater, livestock grazing, deforestation, water impoundment, poor irrigation management, increased soil salinity, and global climate change are the main causes of desertification. The different processes involved in desertification include wind erosion, soil erosion, salinity-alkalinity, and waterlogging. Africa, Asia, Latin America, and the Caribbean are the most regions threatened by desertification. The impacts of desertification include environmental impacts, economic impacts, and poverty and mass migration. A number of methods have been used in order to reduce the rate of desertification. These methods include restoring and fertilizing the land, reforestation,

* E-mail: habes@ksu.edu.sa; Phone: 0096614676195; Fax: 00966-1-4676214.

developing sustainable agricultural practices, and the traditional lifestyles.

1. INTRODUCTION

According to the preamble to the Convention to Combat Desertification (CCD) (UNEP ,1995), desertification is defined in as 'land degradation in arid, semi-arid and dry sub-humid areas resulting from various factors, including climatic variations and human activities'. The two main components of desertification include vegetation degradation and soil degradation. These components are caused by overcultivation, overgrazing, reduction in tree cover, and poor irrigation management. Desertification is generally reversible until land is eventually turned into desert, and can occur anywhere in dry areas and not just on desert fringes (Grainger, 1990a). Desertification reversal is accompanied by increases in the content of soil organic carbon, improvements in the physical and chemical properties of soil, stabilization of shifting sand to produce more typical zonal soil, increases in land productivity and biodiversity, and restoration of the ecological balance (Cheng et al., 2004). Desertification reversal exhibits self-regulating and self-organizing ability in semi-arid environments, but the effect differs among regions (Cui, 2003). The reversal process typically requires 3 to 5 years to take hold in regions with an annual rainfall of around 500 mm, but in regions with annual rainfall below 200 mm, a much longer time could be required (Cui, 1990).

Multispectral and hyperspectral remote sensing data and Geographic Information Systems (GIS) were widely used for monitoring land desertification (Seixas, 2000; Hostert et al. 2001). In addition, spatial heterogeneity and pattern analysis were also used as a sensitivity measure of desertification change. However, research integrating the spatial heterogeneity information directly with spectral information to discriminate desertification type or grade is very scarce. This chapter focuses on causes, impacts, extent, and management of desertification.

2. CAUSES OF DESERTIFICATION

Desertification is a very complex process and its magnitude depends on factors like environmental conditions and human activities. Some other factors can interact to create conditions likely to lead to desertification. These include the movement of refugees during periods of conflict, inappropriate land use or environmental management, specific socio-economic and political factors.

2.1. Environmental Conditions

2.1.1. Arid climate

High evaporative conditions in an arid environment are responsible for a significant loss of vegetation, especially in sandy desert areas. Moreover, high evaporative conditions within adequate irrigation supplies determine the hydrology, land development, and vegetation of the area (Leitchtweiss Institute, 1979).

2.1.2. Sand Erosion and Deposition

Drifting sand and migrating sand dunes in severe wind erosion areas are a continuous threat to potential agricultural land, range plant life, settlements, highways and other areas. The main effect is due to sand deposition rather than erosion, although erosion and transport of sand are the primary factors. Sand accumulation will bury crops and destroy arable land (Hagedorn, 1977). In China, 1.09 million km^2 were affected by shifting sand dunes and drifting sand in many desert districts of which 59% is covered by sand dunes (Chinese Academy of Sciences, 1980).

2.1.3. Resalinization of Agricultural Lands

Irrigation waters contain salts in different amounts and proportions. Resalinization of irrigated agricultural lands results from many factors including arid climate, geology and configuration of the terrain. These factors determine soil properties, land drainage, water and crop management practices (Khatib, 1977). Evapotranspiration in arid areas exceeds rainfall. The countries of the Middle and Near East suffer from salinization problem to a greater extent. The main factors of salinization are irrigation water quality, poor drainage, inadequate irrigation water supplies, low rainfall, initial high soil salinity, adoption of poor soil, water and crop management practices, and inadequate agricultural extension services to the farming community.

2.1.4. Reduction in Soil Moisture

The desert soils normally retain moisture for several weeks after precipitation, whereas ground dryness may cause a desert to persist (Walker and Rowntree, 1977).

2.2. Human Activities

Human activities leading to desertification are mainly related to agriculture. The intensification of human activities brings an increased greenhouse effect, causing global warming. Drylands are likely to be especially vulnerable to rises of temperature during the 21st Century.

2.2.1. Unsustainable Agricultural Practices

Unsustainable agricultural practices include extensive and frequent cropping of agricultural areas, excessive use of fertilizers and pesticides, and shifting cultivation without allowing adequate period of recovery.

2.2.2. Unsustainable Water Management Practices

Poor and inefficient irrigation practices, and over abstraction of ground water, particularly in the coastal regions resulting in saline intrusion into aquifers are some of major unsustainable water management practices which has led to problems of desertification. Overabstraction of groundwater without compensatory recharge has led to depletion of groundwater table.

2.2.3. Overgrazing

Grazing land is estimated to cover about one-third of the land surface of world. Livestock grazing is one of the main causes of degradation in arid regions (Li et al., 2008). The effects of grazing on the plant community and soils are considered destructive because of the reduction of ground cover, productivity and litter accumulation, the destruction of topsoil structure, and compaction of soil as a result of trampling (Milchunas andLauenroth, 1993; Manzano and Navar, 2000). These processes in turn increase soil crusting, reduce infiltration, enhance soil erosion susceptibility and cause a decline in soil fertility (Lavadoetal., 1996; Hiernaux et al., 1999; Yates et al., 2000).

2.2.4. Deforestation

The plant community in arid regions is generally patchy and scarce. Cutting wood for fuel purposes and fires destroy the plant community in the desert. People living in an arid environment cut trees to feed animals and use them as a source of dry wood for sale to the rural people (Grove, 1973).

2.2.5. Increase in Atmospheric Dust

Human activities can also induce desertification through production of airborne dust. Reductions in rainfall can be caused by airborne dust and smoke from fires. Activities including grazing and agricultural cultivation that expose and disrupt topsoil can increase the amount of dust blown into the air. Land clearing activities through burning sends up plumes of smoke. Dust and smoke are characterized by relatively large particle sizes. These larger-sized nuclei have the effect of increasing the threshold for droplet formation in clouds, thereby reducing rainfall (Rosenfeld, 2001).

3. Impacts of Desertification

Desertification has severe financial and societal consequences including direct economic losses, increased health and safety hazards, and decreased agricultural productivity.

3.1. Environmental Impacts

The direct physical consequences of desertification may include an increased frequency of sand and dust storms and increased flooding due to inadequate drainage or poor irrigation practices. This can contribute to the removal of topsoil and vital soil nutrients needed for food production, and bring about a loss of vegetation cover which would otherwise have assisted with the removal of carbon dioxide from the atmosphere for plant photosynthesis.

With the expansion of desertified land, large dust storms were observed more frequently in China. In the 1960s eight large dust-storms were observed, while 20 in 1990s.

In March 2002 the most violent dust-storm observed since the 1990s affected over 1.1 million km^2 of land and 130 million people (Huang and Siegert, 2006), which was considered to be relevant to the rapid desertification caused by human activities and climate change in China (Ye et al., 2000; Wang et al., 2004b).

Desertification causes a drop in biological productivity of land which to a decline in economic productivity. It adversely affects the lives of wild species, domestic animals, agricultural crops and people. Desertification can also initiate regional shifts in climate which may enhance climate changes due to greenhouse gas emissions.

3.2. Economic Impacts

Land degradation is not merely an environmental issue, but has social and economic implications as well (Li, 1998). The World Bank estimates that at the global level, the annual income foregone in the areas affected by desertification amounts to US$ 42 billion each year, while the annual cost of fighting land degradation would cost only US$ 2.4 billion a year. In China, desertification causes a direct economic loss estimated between US$ 2–3 billion annually, while the associated indirect loss is 2–3 times more (Zha and Gao, 1997). In some developing countries, the loss caused by land degradation occupies 1–17% of the gross national product. In some tropical regions, the rate has reached 10%.

3.3. Poverty and Mass Migration

Desertification brings hunger and poverty. People living in areas threatened by desertification are forced to move elsewhere to find other means of livelihood. Mass migration is a major consequence of desertification. From 1997 to 2020, some 60 million people are expected to move from the desertified areas in Sub-Saharan Africa towards Northern Africa and Europe.

4. Extent of Desertification

The United Nations Environmental Programme (UNEP) estimated that 35% of the land surface of earth is at risk, and the livelihoods of 850 million people are directly affected by desertification. 75% of the world's drier lands - 45,000,000 km^2 are affected by desertification, and every year 6,000,000 hectares of agricultural land are lost and become virtual desert. In all, more than 110 countries have dry lands that are potentially threatened by desertification. Africa, Asia and Latin America are the most threatened by desertification.

4.1. Africa

About two-third of Africa is desert or dry lands and land degradation affects at least 485 million people or 65% of the entire African population

.There are extensive agricultural dry lands, almost three quarters of which are already degraded to some degree. Most of the African countries depend heavily on natural resources for subsistence and suffer severe droughts. Desertification in Africa is mainly related to poverty, migration, and food security. About 96% of Egypt is desert and only 4% is inhabited by more than 82 million people, situated mostly in the Nile Valley and the Delta. Egypt is classified as territory susceptible to very high to high desertification sensitivity. One of the main problems facing Egypt is the rapid urban encroachment on its fertile land. Both governmental and private sectors have carried out successful efforts to reclaim desert areas reaching more than one million feddans in the last 35 years. On the contrary, Egypt lost 912,000 feddans of its alluvial fertile land due to urbanization (Abd El Halim et al., 1996). The desertification processes existing in Egypt include urban encroachment on expenses of arable land, wind erosion, water erosion, salinization and water logging. It is estimated that 35% of the land area, about 83,489 km^2, of Ghana is prone to desertification, with the Upper East Region and the eastern part of the Northern Region facing the greatest hazards (UNECA, 2007). About 70% of Ethiopia is reported to be prone to desertification (UNECA, 2007), while in Kenya, around 80% of the land surface is threatened by desertification (UNECA, 2007). Nigeria is reported to be losing 1,355 square miles of rangeland and cropland to desertification each year. This affects each of the 10 northern states of Nigeria (UNECA, 2007). Estimates showed that more than 30% of the land area of Burundi, Rwanda, Burkina Faso, Lesotho and South Africa is severely or very severely degraded (UNECA, 2007).

4.2. Asia

About 1.7 billion hectares in Asia are of arid, semi-arid, and dry sub-humid lands. Degraded areas in Asia include expanding deserts in China, India, Iran, Mongolia and Pakistan, the sand dunes of Syria, the steeply eroded mountain slopes of Nepal, and the deforested and overgrazed highlands of the Lao People's Democratic Republic. Asia is the most severely affected continent with respect to the number of people affected by desertification and drought.

In China, and particularly in west China, the over use and the inadequate use of land has contributed to a number of serious environmental problems, especially desertification. In 1999, China had 2.67 million km^2 of land classified as experiencing some degree of desertification (27.9% of the land area), an increase of 52,000 km^2 compared to 1995 (State forestry administration of China, 2002). With the social-economic development, deserts are expanding and desertification processes have accelerated in North China in the past 50 years (Huang and Siegert, 2006). In this region, sandy desertification driven by wind erosion is one of the main types of desertification (Wang and Zhu, 2003).

The major causes of land degradation in India are deforestation, unsustainable agricultural and water management practices, land use changes for development, and industrialization. The major process of land degradation in India is soil erosion due to water and wind erosion. It contributes to greater than 71% of the land degradation in the country. Soil erosion due to water alone contributes to about 62% and that by wind erosion 10 %. The other processes include problems of water logging, salinity-alkalinity. In Iran more than 85% of the country's 164.8 million hectares is occupied by arid, semi-arid and hyperarid regions with 34 million hectares of desert. The most part of Iran is susceptible to desertification.

Saudi Arabia is a vast desert area representing about four-fifths of the Arabian Peninsula and it comprises 98.5% of hyper-arid to arid lands. The high aridity in Saudi Arabia makes the sand areas that constitute one-third of the total area more serious for agricultural land and urban areas. Desertification in Saudi Arabia could occur due to natural causes or man-made. Much works were carried out in Saudi Arabia to minimize the menace of wind erosion, control of sand dune movement and enhancement of soil moisture conservation (Amin, 2004).

4.3. Latin America and the Caribbean

About one-quarter of Latin America and the Caribbean are actually desert and dry lands. Poverty and pressure on land resources are the main reasons of land degradation in many of these dry areas.

4.4. Other Regions and Countries Affected by Desertification

Most of the countries located in the Northern Mediterranean are semi-arid lands and subject to seasonal droughts. These countries are characterized by high population densities, heavy concentrations of industry, and intensive agriculture. The poor agricultural practices are the major reason of land degradation in these countries. Most of the Central and Eastern Europe is affected by desertification. About 74% of the lands in North America is severely or moderately affected by desertification. About 30% of the land in the United States of America (USA) is affected by desertification.

5. Controlling Desertification

5.1. Restore and Fertilize the Land

To restore degraded lands, crop techniques should be improved by stabilizing the soil while enriching them with organic matter, selecting and associating different crop varieties as in polyculture, and reducing land pressure. A simple and cheap way to fertilize the land is to prepare compost, which will become humus and will regenerate the soil with organic matter.

5.2. Combat the Effects of the Wind

This can be achieved by constructing barriers and stabilizing sand dunes with local plant species. Sand stabilization studies are important for the improvement of soil productivity. Bader (1989) investigated the scientific means to stabilize sand dunes in the eastern region of Saudi Arabia. His results revealed that sand drift and dune movement are the most serious natural problems facing Arabian Peninsula due to vast expansion of cities, roads, industries and agricultural development.

5.3. Reforestation

Trees help fix the soil, act as wind breakers, enhance soil fertility, and help absorb water during heavy rainfall. Because the burning of land and forests increases dangerous greenhouse gases, planting new trees can help reduce the negative impacts of resulting climate change.

5.4. Develop Sustainable Agricultural Practices

Sustainable agriculture refers to the ability of a farm to produce food indefinitely, without causing irreversible damage to ecosystem health. Dry lands are home to a large variety of species, which can also become important commercial products. Agriculture biodiversity must be preserved. Land overexploitation shall be stopped by leaving the soil 'breathe' during a certain-time period, with no cultivation, nor livestock grazing.

5.5. Traditional Lifestyles

Traditional lifestyles as practiced in many arid zones offer examples of harmonious living with the environment. In the past, nomadism was particularly adapted to dry lands conditions. Nomads are herders who migrate thought the year looking for watering holes and pasture for their animals in a way that allows them to utilize the limited resources of their environment over several weeks or months. Changing lifestyles and population growth increase pressure on scarce resources and vulnerable environments.

References

Abd El Halim, M.N., El Mowelhi, M.F., Hawela, H., Kamel, H., El Khattib, H.M., Saleh, H., Nabawi, S., El Akyabi, A., and Ghobrial, K.R. (1996). Remote sensing technique as a tool for detecting environmental changes. *Journal of Soil Science,* Egypt, 36 (14), 289– 305.

Amin, A.A. (2004). The extent of desertification on Saudi Arabia. *Environmental Geology*, 46, 22–31.

Bader, T. (1989) Scientific means and studies used to stabilize dunes in the eastern region, Saudi Arabia. *Proc. Workshop on Desert Studies in the Kingdom of Saudi Arabia.* Center for Desert Studies, King Saud Universty Riyadh. pp 45–66.

Cheng, S.L., Ouyang, H., Niu, H.S., Wang, L., Zhang, F., Gao, J.Q., Tian, Y.Q. (2004). Spatial and temporal dynamics of soil organic carbon in reserved desertification area: a case study in Yulin City, Shanxi Province, China. *Chinese Geographic Science*, 14.245–250.

Chinese Academy of Science, (1980). Combating desertification in China. In: Biswas M, Biswas A (eds) *Desertification; Associated Case Studies for the United Nations Conf. on Desertification*. New York: Pergamon Press pp 109–163.

Cui, W.C. (2003). Application of theory of dissipative structure in reversion process of desertification. *Arid Land Geography*, 26(2):150– 153 (in Chinese with English abstract)

Cui. W.C. (1990). The Malkov's model applied in monitoring desertification process by remote sensing information. In X.P. Yu (Ed), *Study on ecology and environment by remote sensing*. Science Press, Beijing, pp 110–114 (in Chinese).

Grainger, A. (1990a), T*he Threatening Desert: Controlling Desertification*, London, Earthscan Publications.

Grove, A. (1973). Desertification in African environment. In D. Dalby, and R.H. Church RH (Eds), Drought in Africa. London: *School of Orien and Afric Stud,* pp 35–45.

Hagedorn, H. (1977). Dune stabilization. A survey of literature on dune formation and dune stabilization. Eschborn: *German Agen Tech Coop* Ltd. (GTZ)

Hiernaux, P.,Bielders,C.L.,Valentin,C.,Bationo,A., and Fernandez-Rivera,S.(1999). Effects of livestock grazing on physical and chemical properties of sandy soils in Sahelianrange lands. *Journal of Arid Environment,* 41,231–245.

Hostert, P., Roder, A., Jarmer, T., Udelhoven, T., and Hill, J. (2001). The potential of remote sensing and GIS for desertification monitoring and assessment. *Annals of Arid Zone*, 40, 103–140.

Huang, S., and Siegert, F. (2006). Land cover classification optimized to detect areas at risk of desertification in North China based on SPOT VEGETATION imagery. *Journal of Arid Environments*, 67, 308–327.

Khatib, A. (1977). Present and potential salt-affected and water logged areas in the countries of the near east in relation to agriculture. *FAO Irrig Drain* Paper No. 7:13–28.

Lavado, R.S., Sierra, J.O., and Hashimoto, P.N. (1996).Impact of grazing on soil nutrients in aPampean grassland. *Journal of Range Manage*, 49,452–457.

Leichtweiss Institute, (1979). *The water potential of the Al-Hassa Oasis*. Pub. No. 38., Al-Hofuf, Saudi Arabia.

Li, C.,Hao,X.,Zhao,M.,Han,G., and Willms,W.D. (2008).Influence of historic sheep grazing on vegetation and soil properties of a desert steppe in Inner Mongolia. *Agriculture Ecosystem Environment*, 128,109–116.

Li, Y. (1998). Land degradation: its social economic dimensions. *China Environmental Science*, 18, 92–97.

Manzano, M.G., and Navar,J. (2000).Processes of desertification by goats overgrazing in the Tamaulipanthorn scrub (matorral) in north-eastern Mexico. *Journal of Arid Environment*, 44,1–17.

Milchunas, D.G., and Lauenroth,W.K.(1993).Quantitative effects of grazing on vegetation and soils over a global range of environments. *Ecological Monographs,* 63, 328–366.

Rosenfeld, D. (2001). Smoke and desert dust stifle rainfall, contribute to drought and desertification. Arid Lands Newsletter 49. University of Arizona.

Seixas, J. (2000). Assessing heterogeneity from remote sensing images: The case of desertification in southern Portugal', *International Journal of Remote Sensing,* 21, 2645–2663.

State Forestry Administration of China, (2002). '*The Second National Desertification Survey Report (in Chinese)*', Technical Report, State Forestry Administration of China, Beijing.

UNEP, (1995). *United Nations Convention to Combat Desertification*, Geneva, United Nations Environment Programme.

United Nations Economic Commission of Africa (UNECA) (2007). Africa review report on drought and desertification. Available at www. http://www.uneca.org/eca_resources/

Walker, L., and Rowntree, P.R. (1977). The effect of soil moisture on circulation and rainfall in a tropical model. *Quaterly Journal of the Royal Meteorology* ,103,29–46.

Wang, T., and Zhu, Z. (2003). Study on sandy desertification in China: definition of sandy deserfification and its connotation. *Journal of Desert Research,* 23, 209–214 (in Chinese).

Wang, X., Dong, Z., Zhang, J., and Liu, L. (2004b). Modern dust storms in China: an overview. *Journal of Arid Environments*, 58, 559–574.

Yates, C. J., Norton, D.A., and Hobbs, R.J. (2000).Grazing effects on plant cover, soil and micro climate in fragmented wood lands in south-western Australia: implications for restoration. *Australia's Ecology,* 25, 36–47.

Ye, D. Z., Chou, J. F., Liu, J. Y., Zhang, Z. X., Wang, Y. M., and Zhou, Z. J. (2000). Causes of sand-stormy weather in Northern China and control measures. *Acta Geographica Sinica*,55, 513–521.

Zha, Y., and Gao, J. (1997). Characteristics of desertification and its rehabilitation in China. *Journal of Arid Environments*, 37, 419–432.

INDEX

E

F

G

H

I

J

K

L

M

N

O

P

Q

R

S

T

U

V

W